Gems, Colours & Wild Stories

175 years of Constantin Wild

by Nina Hald

with contributions by
Vivienne Becker, Katinka Champion,
Maria Doulton, Richard R. Drucker,
Çiğdem Lüle, Katerina Perez
and Gary Roskin

to all the gems I've loved before...

Gems, Colours & Wild Stories

ERNESTO ARTILLO

175

YEARS OF EXCELLENCE AND UNIQUE EMOTIONS

Table of contents

42
43
50
51

A life with gemstones

by Nina Hald

In 2022, Constantin Wild & Company proudly celebrates 175 years.

With a gemstone heritage spanning more than 400 years of family history, ten generations and nearly two centuries in what my ancestors and I have considered the greatest business of all, I take pride in our company vision: to be the most respected and trusted partner for jewellery brands and designers in the world of exceptional gems.

In 2022, Constantin Wild & Company proudly celebrates 175 years.

Originating from the historic village of Idar-Oberstein in Germany, we are a family enterprise that is regionally based but acts globally. Throughout the centuries we have maintained honest relations with our customers and business partners. We are aware of our responsibilities towards our employees and society itself.

Pursuing gemstones has taken my ancestors and me across the globe many times. During these travels, we have met interesting characters and experienced high drama and world history in monumental landscapes while searching for the most beautiful gems.

Whenever personal contacts give me the opportunity, I travel to the mines where the gems are found. Every gemstone is examined several times before it is bought. I am on first-name terms with each one, because I familiarise myself with its entire life history. It matters a great deal to me that beautiful women can wear my gemstones with a clear conscience.

The foundation of this is love. Our love for gems. These treasures of Nature enchant the eyes and capture the heart of everyone fortunate enough to come into contact with them.

left:
Constantin Wild.

above:
mandarin garnet rough, Namibia,
imperial topaz, 33 carat Brazil.
left:
Casa Conte Mosella.

The success of Constantin Wild & Company is the result of generations of hard work and travel, honouring our promises and keeping our word – every time. As gemstone cutters and merchants, we continue to supply some of the world's most prestigious jewellery companies. On many such occasions we have been given the nickname 'Earl of Colour'.

My ancestors and I have travelled the globe, always striving to be among the first when a new find has been made and to be able to pick and choose from the very best. The discovery of a hue none of us knew existed is a rare revelation, but it does happen once in a while. The superior quality of Constantin Wild gems goes without saying; it is our hallmark, and we take pride in upholding our promise to deliver mastery throughout generations.

Cutting is both a science and an art form – but adding soul to a stone demands experienced intuition and a 'sixth stone sense'. From raw material to luxury item, from the first glance to the final cut, masters of German craftsmanship ensure perfection.

Bathe in an ocean: gems such as paraíba tourmalines, aquamarines and Cambodian zircons come to life with their rich intensity and luxurious range of blue colours. Take a walk in a lush forest: gorgeous shades of green are represented by beryls, tourmalines, tsavorites and demantoids. Tanzanites and imperial topazes captured my heart decades ago. Tanzanites range from blue-blue to violet-blue. The imperial topaz mesmerises the eye with its delicate shades and mix of orange, gold, pinkish-orange, pink and red. Spinels, sapphires, tourmalines and garnets in all their rainbow nuances are a delight to behold.

BLUE TOURMALINE

3 carat

With a family heritage spanning more than ten generations, our mission remains to illuminate the colours of the world's most exciting gems, revealing natural luxury and inspiring unique emotions.

Each gemstone is incomparable. Its type, colour and cut give it its own, very special personality; vivacious or discreet, casual or elegant, conspicuous or subtle, classic or extravagant. Gems summarise all these expressions.

Dealing in precious gems may appear to be a glittering trade. Some imagine the gemstone merchant as a businessman in a dark suit with manicured fingernails. The reality is usually much more mundane. The road from unprepossessing mineral to illustrious gem is long, arduous and generally rather dusty. Never the less, those moments of coming face to face with the majesty of Nature's creation are everything I could wish for.

With this book, I hope to captivate, inform and enchant you with the ultimate gems. It all begins with a selection of the most sublime colours.

Welcome to my world of prestigious beauty.

Constantin Wild

above:
Constantin Wild in the Cruzeiro Mine, Brazil.
left:
Iguazū Falls, Brazil.

PARAIBA TOURMALINE

2 carat

CANARY TOURMALINE

16 carat

AQUAMARINE

22 carat

RUBY

6 carat

DEMANTOID GARNET

5 carat

MANDARIN GARNET

13 carat

VIOLET SAPPHIRE

8 carat

TANZANITE

52 carat

RED TOPAZ

4 carat

SPINEL

5 carat

BLUE ZIRCON

34 carat

TSAVORITE

4 carat

AQUAMARINE

44 carat

PINK TOPAZ

27 carat

IMPERIAL TOPAZ

34 carat

LAGOON TOURMALINE

6 carat

VIOLET SAPPHIRE

5 carat

SPHENE

27 carat

TSAVORITE GARNET

5 carat

IMPERIAL TOPAZ

10 carat

PARAIBA TOURMALINE

3 carat

PINK SAPPHIRE

2 carat

PURPLE SAPPHIRE

5 carat

RED TOPAZ

7 carat

ORANGE SAPPHIRE

10 carat

PARAIBA TOURMALINE

1 carat

CANARY TOURMALINE

26 carat

RED TOPAZ

4 carat

LAGOON TOURMALINE

5 carat

IMPERIAL TOPAZ

37 carat

CHROME TOURMALINE

34 carat

YELLOW SAPPHIRE

27 carat

LAGOON TOURMALINE

12 carat

MORGANITE

24 carat

BLUE ZIRCON

20 carat

DANBURITE

36 carat

PARAIBA TOURMALINE

1 carat

RUBY

4 carat

SAPPHIRE

5 carat

ORANGE SAPPHIRE

13 carat

CANARY TOURMALINE

18 carat

BLUE ZIRCON

38 carat

GREEN TOURMALINE

29 carat

AQUAMARINE

73 carat

IMPERIAL TOPAZ

13 carat

IMPERIAL TOPAZ

75 carat

RED TOPAZ

6 carat

INDIGOLITE

11 carat

WILD
ADVE
NTUR
ES

When Constantin Wild got the very first inklings through the 'gem grapevine' that some interesting green stones had turned up in the heart of Russia, off he went, in the dead of winter – optimistic as only a man on a treasure hunt can be. The gems that prompted the gemmologist and gem merchant to leave at just a few days' notice were otherwise believed to have vanished, all finds exhausted, leaving no trace of their existence. However, the 'call of the Wild' beckoned, an irresistible pull for generations of Wild men, and there was no doubt in Constantin Wild's mind. So, even when the powerful four-wheel-drive military vehicle broke down due to the extreme weather conditions and a bad road, there was nothing he could do but send up a prayer to his deceased father, a gemstone merchant like himself, and hope for help to arrive.

by Nina Hald

"It is with a distinct sense of dread that one gets stuck in snowdrifts several metres high in Russia, knowing that the cold is a real killer," Constantin Wild retells that particular adventure of his, anxiety in his voice even many years later.

As one of the first merchants at the time, Constantin Wild had travelled to Russia, landing at Ekaterinburg Airport in minus 30 degrees Celsius. In an old sedan, he and his contacts trundled off hundreds of miles through the taiga after some incredible gems. Even the prospect of seeing them was quite a thrill: the legendary stones of the tsars, known as demantoids – so named because they were considered 'diamond-like'. They are radiant green, with a mysterious sparkle and fine dark needles, the so-called horsetails that run through the ones found in Siberia, making them unmistakable. These fabulously green stones had been set by the royal court jeweller Carl Fabergé (1846-1920) in creations for the Russian court at the turn of the 20th century, but after the Russian Revolution, they had disappeared from the trade and almost faded from the memory of the gem-stone industry in the decades since.

above:
Accident in the Russian taiga on the way to the demantoid mines.

When he finally arrived at their destination, Constantin Wild was presented with a small bag of rough green pebbles. "It was a calculated risk for sure, but I decided to buy some – and I was the only one to do so. Today, I am proud to say that I made a contribution to the rediscovery of this magnificent stone, and the demantoid – among the finest of the garnets – quickly grew to be a favourite of gem enthusiasts," says Constantin Wild from the luxurious comfort of his Gem Room at the palatial Wild villa in the main street of Idar-Oberstein, the historic gem-cutting centre of Germany.

OF CUSTOMS, COIN POCKETS AND CUTTING DISCS

Crossing borders is always a challenge, and what is true for tourists applies twice as much when you're travelling with gems. "On my first trips to Africa I learned that a hundred-dollar bill can open both doors and borders, help you pass state lines, speed up visa applications and ease emigration and customs procedures on departure tremendously," Constantin Wild explains.

The coin pocket in his five-pocket jeans - his favourite trousers - is also his gem hideaway, where gems worth millions are often transported. "Gems are the densest way to transport high monetary value, and so the safest way to carry them home from my buying trips was to conceal them in the little right-hand pocket of my five-pocket jeans. While all stones were always correctly declared and imported into Germany, the export procedures of some producing countries were so dangerous that it was safer for me to just simply smuggle them out with an innocent smile on my face," says Constantin Wild about his early days in the business. On his first business trips to Moscow and St. Petersburg in the late 1980s, he flew into Russia without applying for a visa. "The process would have taken too long, and if I've learned one thing in the gem business, it is that time is money. As a gemstone merchant you have to reach certain places fast, preferably as the first, especially when a new find has been made - as with the demantoids back in those days. My Russian friends and business partners managed to get me into the country without proper papers a few times, and only once leaving St. Petersburg did the customs people get suspicious of me; luckily Lufthansa, my favourite airline, waited for me, closing the doors as soon as I had boarded the plane," he says.

In April 2004, Constantin Wild had an important order of demantoid garnets to fill for an international client - so when he couldn't get permission to travel to Ekaterinburg, his business partners sent a courier with the gems to meet Wild in Warsaw. "I still remember flying in to Poland, just a week before it became a member of the EU, meeting this gentleman at a hotel in Warsaw - he was very well educated, smart and well dressed, and I felt as if I was in a James Bond situation, exchanging gems for cash. Just as discreetly as he had appeared, the man disappeared again. I always suspected that the courier must have been a KGB officer. The next day, I smuggled the gems out of Poland and into Germany in 'five-pocket-jeans' style, to Frankfurt Airport, where I officially imported them, only to ship them, in official export style, to a faraway country the next day."

Once upon a time, travelling to Brazil, Constantin Wild was asked to take some German-made cutting discs, slicing slabs and polishing powder to his business acquaintances in the country. "On that trip, at Rio customs, even my innocent face didn't help. I was randomly stopped, and in my sports bag was all this heavy sharp equipment, made of metal and with various powders. 'What to do?!', I thought to myself. I decided there and then to give the customs people a long monologue about the latest technical gimmicks and the Walkman that I was carrying, and I must have bored them long enough, because they looked very happy when I left. I left without them even looking through my sports bag, which I managed to carry as though it weighed nothing at all - no easy feat," says Constantin Wild with a laugh.

Constantin Wild has been a 'jeans and worn-out hiking boots' man all his life. He has carried gems worth millions in the tiny fifth pocket in his trousers, through jungles and deserts, only to present them later at fabulous parties, even gaining a certain amount of knowledge of whose necks, wrists and fingers they would end up gracing. Some of the stories can be retold and documented – others cannot for reasons of discretion, and because some details have been lost in the fog of history. The Wild family archives contain treasures such as an old letter from a Russian jeweller named Ivanoff Frères, who from French exile wrote about a collaboration between Carl Fabergé and Johann K. Wild. Old travel documents and passports nestle with distinction among diplomas from the large World Exhibitions from the end of the 19th century in the glass case at Constantin Wild's office. Antique scales are kept alongside family portraits and gemstone specimen collections, demonstrating how Idar-Oberstein has gone from local to global, also due to the contributions of Constantin Wild's ancestors.

The safe in the fifth pocket.

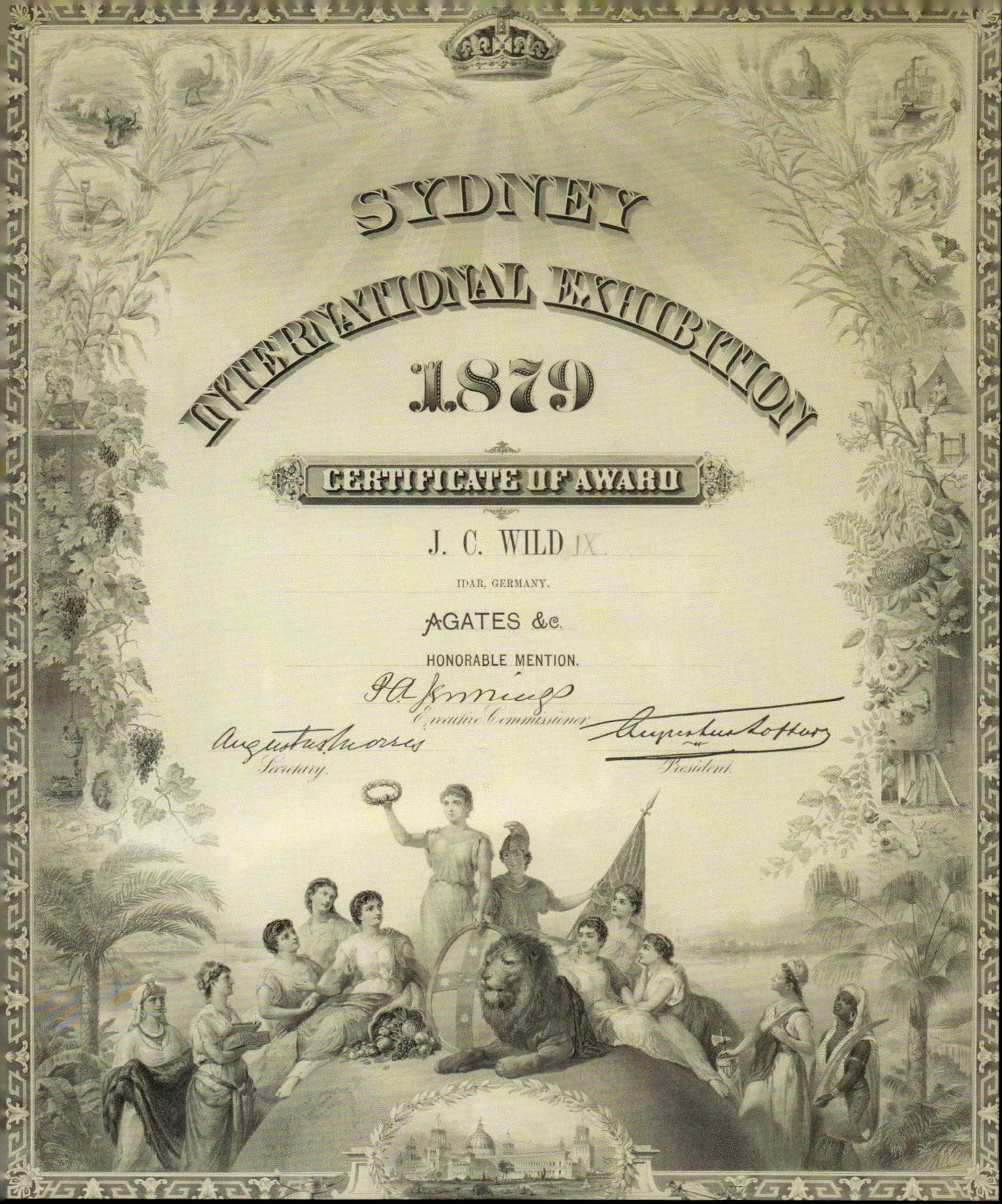
SYDNEY
INTERNATIONAL EXHIBITION
1879
CERTIFICATE OF AWARD
J. C. WILD IX.
IDAR, GERMANY.
AGATES &c.
HONORABLE MENTION.
Executive Commissioner
Secretary.
President.

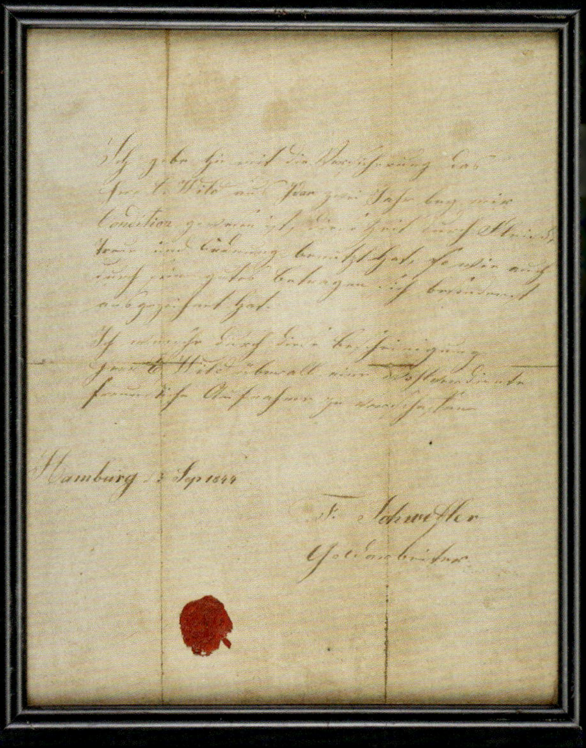
Hamburg
F. Schwetler

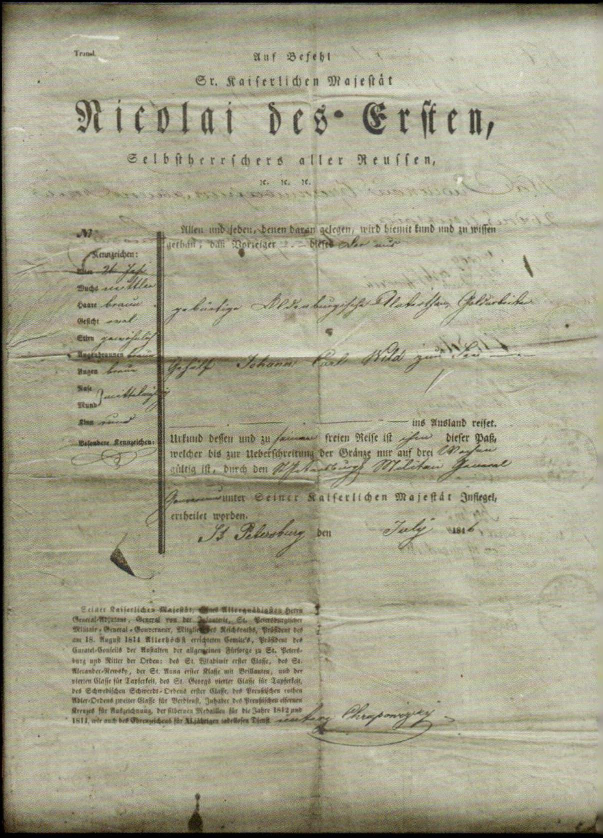
Auf Befehl
Sr. Kaiserlichen Majestät
Nicolai des Ersten,
Selbstherrschers aller Reussen,
Allen und Jeden, denen daran gelegen, wird hiemit kund und zu wissen gethan, daß Vorzeiger … Johann Carl Wild … ins Ausland reiset.
Urkund dessen und zu … freien Reise ist … dieser Paß, welcher bis zur Ueberschreitung der Gränze nur auf drei … gültig ist, durch den … unter Seiner Kaiserlichen Majestät Insiegel, ertheilet worden.
St. Petersburg den … Juli 1846

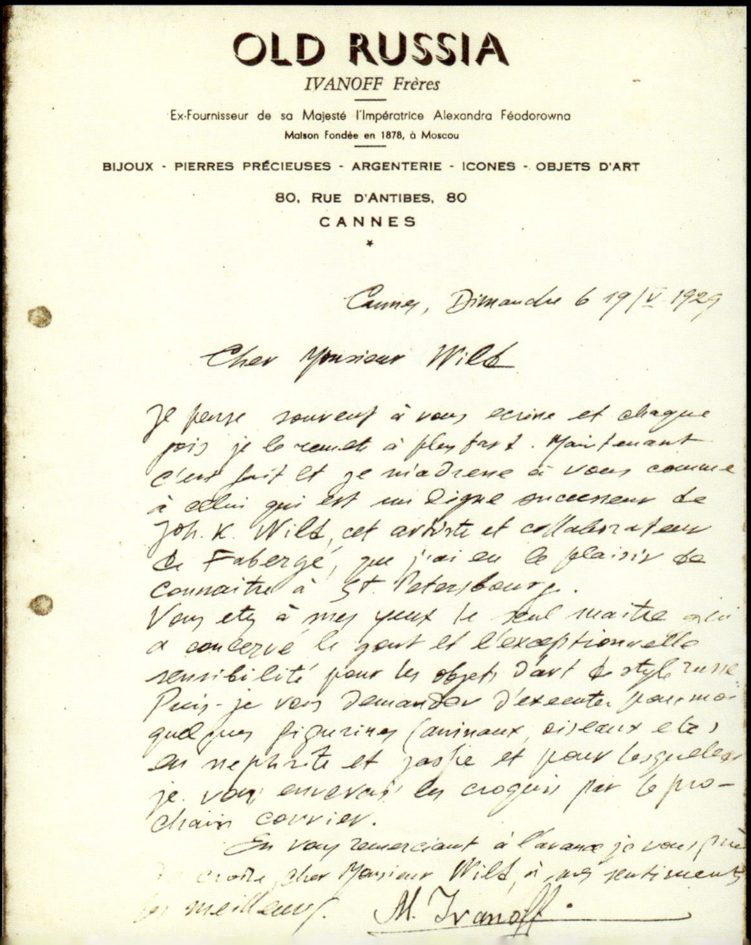
OLD RUSSIA
IVANOFF Frères
Ex-Fournisseur de sa Majesté l'Impératrice Alexandra Féodorowna
Maison Fondée en 1878, à Moscou
BIJOUX - PIERRES PRÉCIEUSES - ARGENTERIE - ICONES - OBJETS D'ART
80, RUE D'ANTIBES, 80
CANNES

Cannes, Dimanche 6 19/V 1929

Cher Monsieur Wild

Je pense souvent à vous écrire et chaque fois je le remets à plus tard. Maintenant c'est fait et je m'adresse à vous comme à celui qui est un digne successeur de Joh. K. Wild, cet artiste et collaborateur de Fabergé, que j'ai eu le plaisir de connaître à St. Petersbourg.
Vous êtes à mes yeux le seul maître qui a conservé le goût et l'exceptionnelle sensibilité pour les objets d'art & style russe.
Puis-je vous demander d'exécuter pour moi quelques figurines (animaux, oiseaux etc.) en néphrite et jaspe et pour lesquelles je vous enverrai les croquis par le prochain courrier.
En vous remerciant à l'avance je vous prie de croire, cher Monsieur Wild, à mes sentiments les meilleurs.

M. Ivanoff

clockwise:
Certificate of award for Johann Carl Wild IX exhibiting agates and haematites at the 1879 Sydney International Exhibition.
Certificate of employment of Schwetler for Johann Carl Wild IX working as a goldsmith in Hamburg in 1844.
Travel document, St. Petersburg 1846.
Letter from Ivanoff Frères, Cannes 1929.

clockwise:
Historic view of Oberstein.
Agate cutter at work.
Rough gemstone auction in the front yard of W. Constantin Wild & Co., c.1920s.
Sandstone wheels lined up to be used in an agate cutting mill.
Idar Hauptstrasse c.1910.
Miners at work at W. Constantin Wild's mine in Nunkirchen.

The Wild company's headquarters are located at the historical centre of Idar-Oberstein on the Main Street (Hauptstrasse in German) opposite the German Gemstone Museum which, housed in an old pearl merchant's villa, exhibits an impressive collection of rough, cut and carved gemstones. Furthermore, the headquarters are surrounded by such quaint establishments as the Diamond Apothecary, the Diamond Bourse and the Opal Hotel, with a view up to the mediaeval castle and the famous 'Church in the Rock', or *Felsenkirche*.

Named after the Idar stream that supplied energy to cutting mills for centuries, and Oberstein after the 'Lords of Stones', the aristocratic rulers of the time – Die Herren von Stein, first named in the year 1075 – the two small villages were bustling centres of activity for hundreds of years. The inhabitants of Idar provided the cut gemstones to the metalworkers and goldsmiths of Oberstein. Of course, rivalries were always hot topics for gossip between the two villages, and in January 1820, the talk of Idar was the illegitimate son of Anna Eva née Wild (1786-1851). She herself had been born into one of the Wild gemstone-cutting family dynasties and had wed goldsmith Johann Carl Bohrer (1786-1815). As a widow, 34 years of age, she fell in love with a 20-year-old goldsmith's journeyman named Johann Carl Werle, descended from the minstrels of Strasbourg, and bore him a son in 1820 – but he married someone else. In those days, a child out of wedlock was considered a sin, but the family gathered around the boy, who was baptised Johann Carl Wild IV (1820-1893), just four days after his birth – no expenses spared. What the child lacked in parentage on his father's side – officially the father was registered as unknown – he had in abundance on his mother's side. Anna Eva Wild's ancestors had lived as agate cutters and goldsmiths for at least five generations in Idar, and his first godfather was goldsmith Johann Wilhelm Wild (1759-1837), an uncle of the child's mother. Other godparents were the goldsmith Johann Jacob Wild (1784-1862) and his wife, Maria Elisabetha née Wild (1786-1856), who was a cousin of the child's mother.

"The surname Wild is among those most frequently encountered in Idar-Oberstein. It is borne by several gem merchants and cutters, and in one way or another, all of us are related. Members of my family were probably among the very first gemstone cutters in Idar. Our coat of arms bears the year 1557, and without a break my family traces its roots back to the birth of Johann Nicol Wild in 1673, who by the bye was one of the founders of the very powerful goldsmiths' guild," Constantin Wild retells the family history.

GEOGRAPHY & GENETICS

Located at the south-western edge of Germany in the Rhineland-Palatinate towards the border of France, between the ancient cities of Strasbourg and Frankfurt and the World Cultural Heritage sites of Heidelberg and Trier, Idar-Oberstein is home to the historic stone-cutting centre of Europe.

It all began some 600 years ago with a real stroke of luck, when the first large agate deposits were found there at the end of the 15th century. The finds resulted in regular production that - from at least 1454 - supplemented the farmers' livelihood. Idar housed the gem cutter; in Oberstein, goldsmiths set the cut gems in jewellery. A discovery of agates, jasper, amethyst and rock crystals in the nearby Hunsrück further cemented the cutting industry. From 1834 onwards, the industry flourished when German emigrants to Brazil, including members of the Wild family, sent agates, jasper, rock crystals and amethyst geodes back to Idar as ballast in sailing vessels to be processed. Since the 1870s, coloured gemstones from all over the world have made their way to Idar for processing and trade.

Some 55 of the 157 water-powered mills built on both sides of the Idar stream made cutting and polishing of agates possible at minimal cost, provided it was not a drought year. The 'real cost', however, was in human lives. The average cutter usually had a life span of about 40 years, dying from some terrible lung disease. While Johann Wolfgang von Goethe (1749-1832) termed water "a friendly element to whomever is familiar with it and knows how to treat it" - and it was indeed water that powered the huge sandstone grinding wheels - it also made for a damp work environment in which cutters literally got 'stone lungs' from inhaling the dust.

As was traditional at the time, fathers would teach sons, passing down cutting techniques from one generation to another. Idar was built on centuries of hard work, diligence and skilled craftsmanship, and to this day, Idar-Oberstein is the only gemstone centre in the world where coloured gemstones of all kinds are processed and traded. Knowledge amassed over centuries and thorough German training produce highly qualified craftsmen who do not merely command cutting; they also know how to judge and treat raw stones properly and have the creativity and skill to transform complex stones into extravagant works of art.

THE WILD FAMILY TREE

Caroline Wild
1849 - 1898

Bertha Wild
1853 - 1886

Johanna Wild
1855 - 1873

Clara Wild
1857 - 1860

Emilie Luise Wild
1859 - 1929

Eva Auguste Wild
1861 - 1935

Charlotte Wilhelmine Wild
1863 - 1941

Carl-August Wild **1851 - 1911**
Pearl Dealer

Georg O. Wild **1892 - 1975**
Founder of the German Gemmological Association

Klaus Eberhard Wild ***1929**
Gem Merchant

Johann Carl Wild IX
'Russ-Carl'
1820 - 1893
Company Founder
Gem Merchant and Goldsmith
in Hanover, Hamburg & St. Petersburg
⚭ Caroline Veeck

Wilhelm Constantin Wild
'Conte'
1866 - 1956
Gem Merchant
⚭ Wilhelmine Darstein

Anna Eva Wild **1786 - 1851**
⚭ Johann Carl Werle
Goldsmith

Johann Carl Wild **1749 - 1813**
Goldsmith

Johannes Wild **1711 - 1787**
Founder of the Goldsmiths' Guild

Johann Nicol Wild **1673 - 1720**
Agate Cutter

Johannes Wild **1652, ref. 1676**
Agate Cutter

Johannes Wild **ref. 1639 - 1640**
Agate Cutter

Johannes Wild **ref. 1597**
Miller

Johann Carl Heinrich Wild
*2011

Anna Luise Katharina Wild
*2004

Marie Kristin Elisabeth Wild
*2001

Christian Max Friedrich Wild
*1999

Carl Friedrich Ernst Christian Wild
*1968

Julian Wild
*2011

Carmen Wild
*2005

Armin Wild
*2004

Carl Ulrich Alexander Wild
*1966

Caspar Friedrich Maximilian Nagy
*1996

Johann Friedrich Constantin Wild 'Conte'
Gem Merchant
*1964

Joh. Friedrich Carl Wild 'Fritz'
1905 - 1984
Gem Merchant
⚭ **Christel Wild**
1937 - 2008
Gem Merchant

Helma Stolterfoth
1912 - 2001

Peter H. Stolterfoth
*1940

Axel Stolterfoth
*1946

Ellie Wild
1902 - 1982

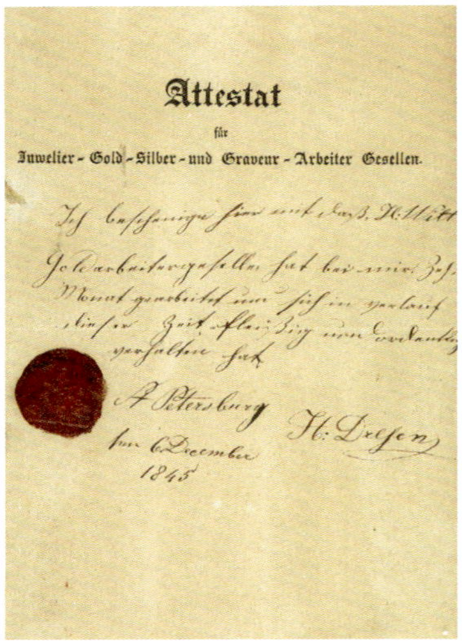
Attestat
für
Juwelier- Gold- Silber- und Graveur- Arbeiter Gesellen.

Certificate for Johann Carl Wild IX as a goldsmith at H. Drehen, St. Petersburg 1845.

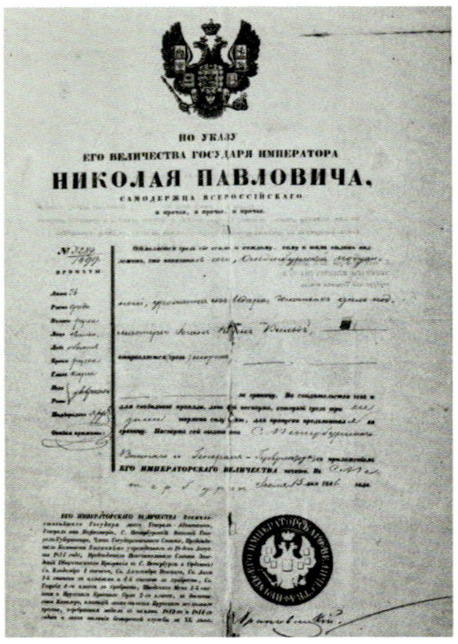
ПО УКАЗУ
ЕГО ВЕЛИЧЕСТВА ГОСУДАРЯ ИМПЕРАТОРА
НИКОЛАЯ ПАВЛОВИЧА,

Travel document of C.F. Herold, St. Petersburg 1846.

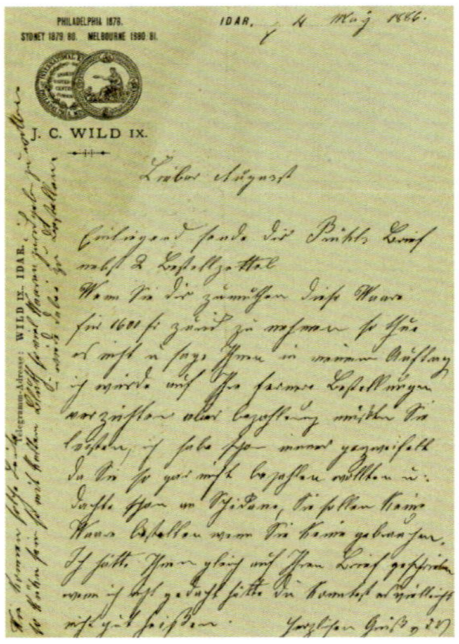
J. C. WILD IX.

Letter from Carl Wild IX to his son, Carl August Wild, 1886.

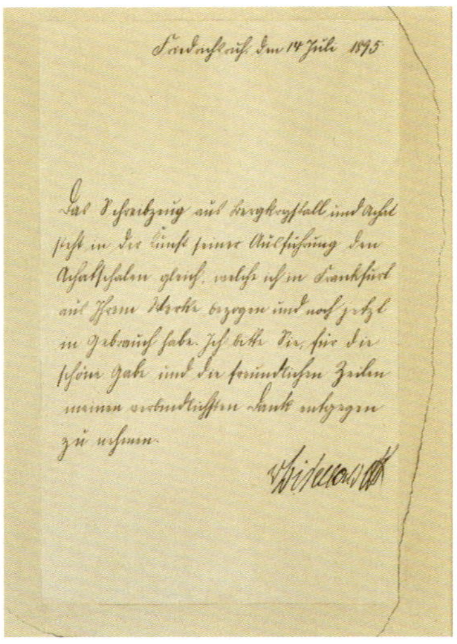

Letter of appreciation to Johann Carl Wild IX from Chancellor Otto von Bismarck, Friedrichsruhe 1895.

As was customary in Europe at the time, Johann Carl Wild learned his mother's family trade and became a goldsmith. Raised lovingly under the matriarchal wing, he developed into a man who was every bit as adventurous as he was ambitious and enterprising. There is firm evidence that young Johann Carl, hav-ing finished his apprenticeship as a goldsmith, worked in far-off Hamburg and Hanover. In 1845-1846, he betook himself – and what a journey it must have been! – to the Russian capital St. Petersburg, where he worked as a so-called 'goldworker's mate'. At the time, St. Petersburg was the centre of the goldsmith arts, and his stay must have been educational in every way. However, by the end of 1846, Johann Carl Wild chose to return to Idar, where he founded his own company – J.C. Wild IX, Idar – in 1847. To begin with, the company processed precious metals as a jewellery house, and soon he took on two journeymen and an apprentice in the workshop he built in the house at Hauptstrasse 105. However, it was not until Johann Carl Wild began importing Brazilian agates and other rough gems that business prospered. With that impetus, he turned to gemstone cutting and trading, like some of his ancestors. In 1849, he wed Caroline Veeck (1825-1895), with whom he had nine children. She was a force to be reckoned with in the company, always helping out, thus also reinforcing the established traditions of the time for spouses pitching in with the running of the family company. Besides the gem business, Johann Carl Wild also kept up the family farm, which was typical of that time, in order to make sure that he could always feed his family.

Shortly after returning home to Idar, Johann Carl Wild had been dubbed 'Russ-Carl' on account of his time spent working in Russia. That did not suit him at all, so in order to distinguish himself from the many other bearers of the name Wild in Idar, he went to the parish register, counted the generations that had come before him and began calling himself Johann Carl Wild IX. It was in this name (and number) that he founded his business. He was so successful in building it – and his name – that he established trade relations with England, India, Ceylon (today Sri Lanka), Hong Kong, Australia, South Africa and America in the 1870s, as reflected by the handwritten order books kept in the company records. So industriously did he commit himself to the business that he also managed to exhibit – and receive awards for excellence – at the world exhibitions in Philadelphia in 1876 and in Sydney in 1879 and Melbourne in 1880. As something of a cosmopolitan for his time, Wild IX found it a must to set up the Paris office; it paid off very well, with the company selling not only to Parisian firms, but from there also to the Swiss, Italian and American markets. Wild IX's industriousness manifested itself in many ways; he sent an exquisite set of 'writing materials made of rock crystal and agate' to Federal Chancellor Otto von Bismarck (1815-1898); alas, the Chancellor's personal letter expressing his thanks for the set – naming the rock crystal and agate – did not arrive until after the death of Wild IX. However, the letter, dated in 1895, has been preserved in the company archives.

right:
Johann Carl Wild IX.

CONTINENTAL

Not until 1935 did business normalise in general, but when it did, W. Constantin Wild & Co. began shipping gems to its traditional export markets such as England and France. In 1938, Fritz Wild went on a business trip to America, visiting New York, Washington D.C., Chicago and San Francisco, but just one year later, in 1939, the advent of World War II (1939-1945) brought the gemstone business to a complete stop, leading to another abrupt decline similar to the one caused by World War I. Once again, all assets, residential and office premises, outstanding debts and warehouses were confiscated. The company's inventory of gemstones was also confiscated and mysteriously disappeared – Fritz Wild tried for the rest of his life to get compensation from the German government but never succeeded – and in addition to the gems, all lists of customers and other business documents also vanished.

"At the onset of World War II, my grandfather Conte Wild picked his best gems, put them in a box and hid it in the basement behind a double wall. When in March 1945, shortly before the end of the war, our premises were confiscated; my grandfather – 80 years old at the time – was given two hours to leave his house. When he was allowed to come back in 1946, the walls had been demolished and everything stolen," Constantin Wild says.

In a small room in 1946, Fritz Wild restarted the business with four loyal employees – once again from scratch. Only thanks to the post-war *Wirtschaftswunder* years, the currency reform in 1948 and decades of hard work did the company gradually improve and regain its strength. Fritz Wild profusely thanked all the employees, the generations of cutters, carvers and jewellers who had supported the family business in good times as well as bad – and of course his wife, Christel, mother of his three sons, for being his rock – at the grand 125-year company anniversary in 1972. Two employees stayed in the company for more than 40 years – two more were honoured with the *Bundesverdienstkreuz* for over 50 years of loyal work.

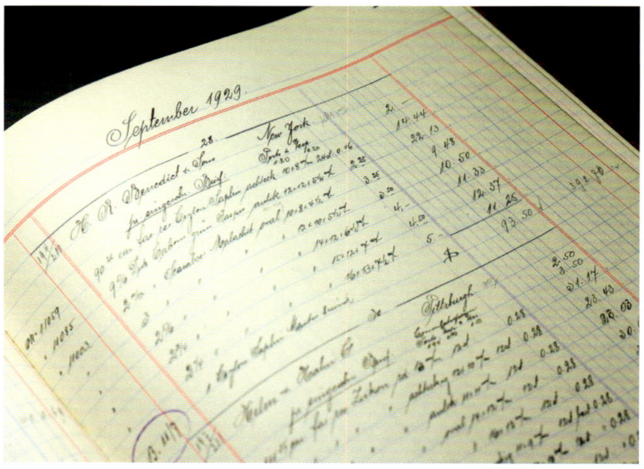

above:
Sales book from 1929.
Stamped inflation money, 1922.

right:
Cut amethyst in a pile of coal.

WASHINGTON D.C. FR

Fritz Wild (first row, second from right) in front of the Capitol, Washington D.C.

When the pretty Christel Vogt (1937-2008) began as an apprentice at W. Constantin Wild & Co., Fritz Wild, aged 55 at the time, immediately took a shine to the young lady, who was only 23. He came from a wealthy family and had travelled the world, studied and worked abroad, and lived and worked in Paris between the wars – and she was swept off her feet by his attentions. Fritz Wild and his friend Max Hahn were referred to as 'the two dream bachelors' in a German magazine article. They both worked with gemstones. Fritz Wild was a true fairy-tale prince, as he owned the family castles bought in the early 1930s, and they both received baskets of love letters from women who wanted to 'marry the dream'. However, no woman succeeded in winning Fritz Wild's heart until Christel Vogt. After a very brief courtship, they became a couple, but as she was so young, he suggested that she spend a year in England as an au pair and attend a housekeeping school in Lausanne, Switzerland, before they married in 1962, and she duly did so. In January 1964, they welcomed their first son, Johann Friedrich Constantin Wild, known as Constantin, the current owner of the family company. Since then, Constantin Wild has spent a lot of time and effort ensuring that the ruins of one of the family castles are safe and preserved for posterity. He has even bought a nearby winery as a hobby project – although he does take his wine appreciation seriously.

"My mother was always working full time in the company, besides raising us children, running the household with maids, and she was the hostess at the dinner parties for business associates when they visited us in Idar-Oberstein. She was a strong, caring, responsible mother and businesswoman, always working by my father's side," Constantin Wild describes his mother.

From 1954 to 1983, the company experienced a great upturn during the course of the economic miracle. Coloured gemstones went to renowned jewellery companies in the Federal Republic, Western Europe and America, and the range was complemented by animal engravings in stone, cameos and bead necklaces. In the 1970s, Christel Wild also travelled alone to visit customers abroad if necessary.

above:
Christel Wild, 1972.
Christel Wild and Robert Lightstein at Gump's dinner reception in San Francisco.
Max Hahn and Fritz Wild in Rio de Janeiro.

next page:
Article about the 'dream castle' on the Rhine river in the German magazine *Ihre Freundin*, January 1950.

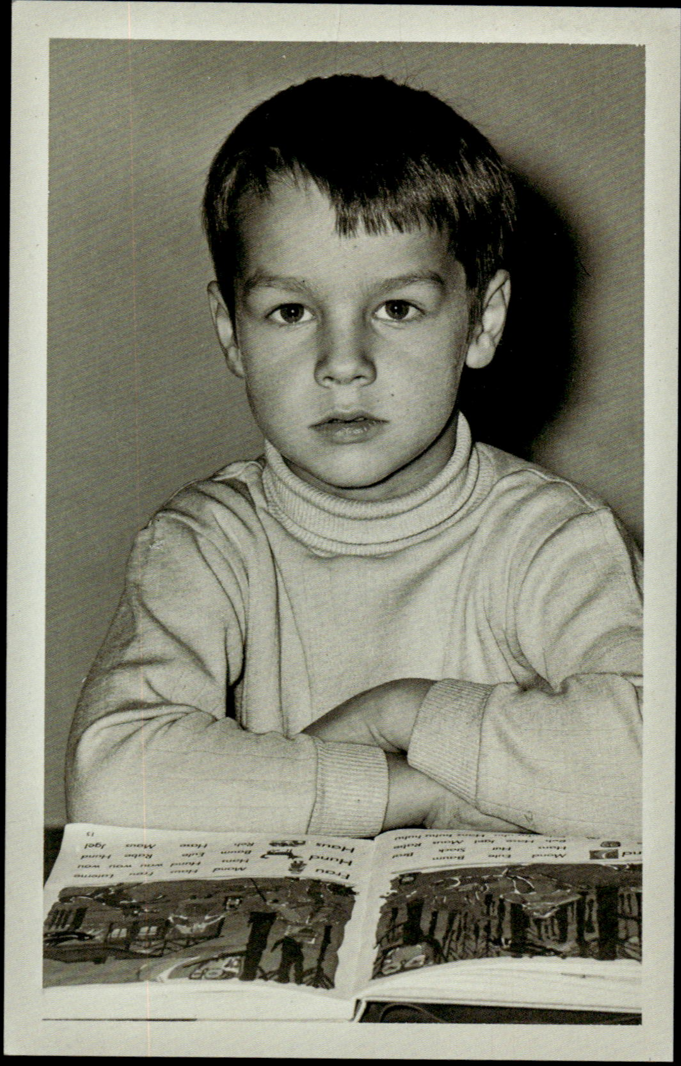

clockwise:
Fritz Wild and Christel Wild at the 125th company anniversary reception at the Park Hotel, Idar-Oberstein 1972.
Christel Wild (left) in the office as an apprentice.
Constantin Wild at primary school.
The Wild Family, 1974.
The Wild boys Alexander, Christian and Constantin.

Ein Gästebuch gibt es natürlich auch in der Burg des modernen Ritters, zu dessen Ehre wir gerne die Kunde verbreiten, daß es ein liebenswürdiger Zeitgenosse mit romantischem Einschlag ist. Das Gästebuch ist in dickes Schweinsleder gebunden, und selbstverständlich trägt der Besucher von heute sich zünftig mit einem Federkiel in die Erinnerung des freundlichen Gastgebers ein, dessen Burg ihn für Stunden der Unrast des Lebens entrückte.

11

THE IMPERIAL TOUCH AND A PERSONAL FAVOURITE

In 1735, topazes with red-orange-shocking-pink-brownish hues and a golden shine found in Brazil were so beautiful that the epithet 'imperial' seemed appropriate. Today, they are still some of the most valuable stones to come from the area, and just like the Burmese imperial jade, they get their colour from the rare element chromium (Cr3+). Imperial topazes have been a personal favourite of both Fritz Wild and his son Constantin Wild. A few years back, a particularly wonderful four-colour imperial topaz of some 28 carats in Constantin Wild's possession made such an impression on even the staunchest 'diamond believer' in the industry that he tried to buy it from him. This would-be purchaser had founded his own immensely successful diamond jewellery company in London, building his professional reputation almost exclusively on jewellery set with diamonds and fancy coloured diamonds. However, Constantin Wild would not part with the topaz at the time

IMPERIAL TOPAZ

34 carat

IMPERIAL TOPAZ

15 carat

IMPERIAL TOPAZ

37 carat

IMPERIAL TOPAZ

20 carat

At the beginning of the 20th century, W. Constantin Wild & Co. had done business with Japanese and Chinese companies. Thus, when an American colleague recommended that Constantin Wild go to Japan in the mid-1990s, since Japan was attracting a lot of attention as the fastest expanding market for coloured gems at the time, he knew from old company ledgers and customers' business cards that it was solid advice to follow up on. "I got on a plane and flew to Tokyo. Knowing no Japanese, I asked the Chamber of Commerce to recommend me a translator. With her I sat all day in my small hotel room, calling jewellers from the yellow pages. At the end of the first work day, the translator all but collapsed, crying, as she had not been able to set up even a single appointment for me. The next day, when we went at it again, she realised that, by mistake, she had translated coloured gemstones as 'colourful glass beads'. No wonder there wasn't a single jeweller who wanted to meet me," says Constantin Wild.

On the same trip, Constantin Wild travelled to Kofu in the Yamanashi prefecture at the foot of Mount Fuji, two hours west of Tokyo by train. For centuries, Kofu had been the jewellery manufacturing centre of Japan, and W. Constantin Wild & Co. had a client there – Yoshiaki Shinohara – who helped Constantin Wild by introducing him to the local jewellery manufacturers. "Thanks to the Japanese jeweller Chino I was invited to attend a Karl Lagerfeld event during one Baselworld fair, and at that dinner, I met Miss Sonja Klettke. Her mother was Japanese, her father German, and her sisters lived in Pforzheim, working as sales agents for German jewellery brands selling to Japan, and so I hired her as my interpreter, later travelling to Japan with her. Thanks to her, I got to understand the Japanese system of import, wholesalers and retailers," says Constantin Wild. Shortly after that, and thanks to the good quality of his gems, Constantin Wild became the gemstone supplier to Seiko's and Citizen's jewellery lines. "On a trip to Japan, I would sell perhaps ten times as much as I could on a trip to America or Switzerland. The Japanese sales helped me tremendously in the revitalisation of the company, allowing me to reinvest," he says.

"Thanks to the Japanese jeweller Chino I was invited to attend a Karl Lagerfeld event during one Baselworld fair."

right:
ICA congress delegates, Tokyo 1995.
Professor Gübelin at Constantin Wild's booth at the Japan Jewellery Fair, Tokyo.

INTERNATIONAL COLORED GEMSTONE ASSOCIATION
CONGRESS 1995 TOKYO

1847

DEMANTOID

1 carat

DEMANTOID

2 carat

Stand at Intergem Fair, Idar-Oberstein.

During Fritz Wild's time, gemstone buyers from Milan, Paris, New York and San Francisco had come to Idar-Oberstein to buy lots from W. Constantin Wild & Co., all laid out on tables. Production parcels would arrive on Fridays, only to be sold the week after. "When these buyers no longer came to Idar-Oberstein, I knew I had to start travelling to them – and I also wanted to exhibit at trade fairs to secure new clients. When I began work in the company, we began exhibiting first in Munich, then in Basel, and later on in Hong Kong, New York, Tucson, Dubai, Tokyo, Taipei and Yokohama. After 1996, I cut back a bit, 'only' exhibiting six times a year; at Baselworld, Tucson, three times in Hong Kong, and the local Intergem show in Idar-Oberstein," Constantin Wild recalls.

In 1996, Constantin Wild arrived at one of the most important decisions of his life. "I wanted my future to be in the company, which was owned solely by my mother at the time, and so I offered to buy it and pay off my two brothers. I was 32 years old; I wanted to organise my own life and get a perspective on what I was doing. This was a big step and a serious risk for me. At that time, I had to borrow all the money from the bank, which added a lot of pressure on me to keep the initial momentum and success of the company going. My plan was to pay back the loans within ten years and, also thanks to lucky circumstances, I made it happen. The rediscovery of the Russian demantoid garnets played a huge part in that," he says.

"The rediscovery of the Russian demantoid garnets played a huge part in that."

Constantin Wild

TRAVELLING THE WORLD

Constantin Wild has enjoyed his fair share of travelling - to New York, Rio de Janeiro, Tokyo, Australia and all the ensuing adventures there: riding a motorcycle from Idar-Oberstein to Moscow and St. Petersburg in the 1980's, going on a somewhat dangerous demantoid buying trip to the Urals before the Iron Curtain fell, eating snakes in Taiwan and insects in Burma, marvelling at the wonders of the world and getting to see some of the most fantastic treasures at the same time. "As one of the first Westerners, I got to visit Mogok in Burma - but all the time, I was escorted by a pickup truck with no fewer than 18 heavily armed soldiers," Constantin Wild says, still wondering if the soldiers were there for his protection or that of the country's population.

"Back in those days, 'cash was king' if you wanted to buy gems in their country of origin. It was always highly risky to travel in with cash and out with gems, and there were usually practical handling hassles when converting currencies. On my trips to Russia in the late 1980s, the official exchange rate was three German marks for one Russian rouble. The street market rate was three roubles for one mark. Needless to say, I used the street guys, but I believe I also have to count myself lucky to be alive," says Constantin Wild.

The adventures have been many and varied. Back in those days, Japan was not an easy country in which to travel. One day at Shinjuku Station, on his way to Kofu, the gem and jewellery capital of Japan in the prefecture of Yamanashi, Constantin Wild found himself at the busy Shibuya crossing, where ten lanes were crossed by six walkways, 2,500 people crossing at each green light, assisting the more than three million passengers changing trains there every day. The Japanese were filtering through here in discipline and respect as usual - until one German gemstone merchant, carrying two heavy shoulder bags and pushing two trolleys, tried to cross the road, stumbling in the middle of the intersection and then lying there, flat on his face, with all his luggage scattered around him and cars approaching from six sides ...! Once, Constantin Wild accompanied his mother to San Francisco to a black-tie party at the luxury department store Gumps, where Marilu Klar and company president Robert Leitstein were hosting an annual party. "My mother was always so proud to take part in all that glamour, and years later she took me to the same party. I remember how it felt that first time, getting all dressed up - in a rented tuxedo, but nevertheless," Constantin Wild recalls.

The first time Constantin Wild went to Russia, he was 23 years old – the year was 1987. "I grabbed the opportunity to be in the first European group allowed to enter Russia. We went on motorbikes; I rode all the way from Idar-Oberstein to Vienna, Bucharest, Lemberg (Lvov), Minsk, Moscow and St. Petersburg. It was a great experience, but I didn't buy any stones on the trip. I got back home again and continued work – until one day in 1988, when a junky old Russian car stopped in front of my company in Idar-Oberstein. Two Russian men stepped out and rang the doorbell. Warily, we admitted them to the premises. Then, standing in reception, they showed me the most gorgeous green garnets, recently rediscovered in the Ural Mountains. Recognising them for what they were from my studies at the GIA – remembering and seeing that the tales of their fabled beauty, high brilliancy, diamond-like lustre and dispersion were indeed true, and finally fully realising why they were so very much in fashion in Fabergé's time – I looked them over for the horsetail inclusions, caused by chrysotile, a type of asbestos. I decided to buy them, and I sold them immediately for a good profit on my next sales trip. That made me follow up, even flying to St. Petersburg during a very cold Russian winter, which was rather reckless of me," Constantin Wild recalls.

From then on, travelling to faraway countries to visit mines became part of Constantin Wild's work life. Some of the trips were rather dangerous due to the soldiers of military regimes not regarding foreigners with a kindly eye; some were perilous due to mines almost collapsing on top of his head – though, he adds on a side note, "Shoddy electrical wires and polluted water pipes were more dangerous than the actual mine constructions, which were usually pretty sturdy on an average basis." Some trips were risky due to natural disasters happening just before or right after Constantin Wild's visit to the country in question. In 1991, he was one of the first Europeans to visit Burma, known today as Myanmar, in order to see the Mogok ruby mines, the country's treasure chamber.

DEMANTOID GARNET
FROM RUSSIA

5 carat

clockwise:
Burmese man, from a photograph taken in Mandalay, Myanmar.
Blue sapphire rough, 1991.
World map of W. Constantin Wild & Co.
Rough stone resources.
Tourmaline mine impressions, Minas Gerais, Brazil 2011.

“FAIR TRADE IS IMPORTANT TO ME.”

Constantin Wild

Sourcing trips to Kenya, Tanzania, Mozambique, Brazil, Burma, Sri Lanka, Russia, Thailand and Australia afford Constantin Wild the opportunity to forge personal contacts with miners and allow him to keep a discreet eye on working conditions. "Fair trade is important to me. I want to make sure that the people working the mines are able to lead healthy lives, that they are properly protected by the safety installations and that they earn a fair wage. I don't just examine every gemstone I buy several times; you could say that I am on first-name terms with it, familiar with its life history, because I see where it came from, and because it means a lot to me that beautiful women are able to wear my gems with a clear conscience. I want to see with my own eyes that people are treated well," he adds.

A good gem merchant relies on his or her sixth sense. When looking at a rough gem, the merchant needs to be able to assess what is hidden beneath the dull surface of an unassuming stone. Only experience and instinct help here, because each coloured stone is as unique as the desires of those on the lookout for something special – valuable gems for sophisticated collections or particularly brilliant investments. "I have lived surrounded by fine gems since the day I was born. I have been in the business myself for close to 40 years – and to this day, I am still fascinated by the almost boundless diversity of coloured gemstones. I am constantly on the lookout for new deposits," says Constantin Wild.

There are coloured gems which really are very special. Among them are new discoveries found only in a single place. Some deposits are extremely small; others may hold only a few individual specimens of gem quality or only small ones. Some have been known and coveted for a long time – but supplies from the mines are exhausted. All these coloured gemstones have one thing in common: they are very rare – much rarer than diamonds. The new rarities include paraíba and canary tourmalines and mandarin garnets, each with fascinating colours. The imperial topaz is a classic but hardly found anywhere at all now. The Russian demantoid is a rediscovered gem, but unfortunately there are only very few crystals of more than half a carat. "I have handled some outstanding individual gems – and I have gathered the ones I have been most proud of in a 'club' that I have named my More Than Just Rare collection," says Constantin Wild. Selling a particularly spectacular 75-carat octagonal-cut multi-coloured imperial topaz gave him the inspiration to do his Gemstone Portraits and institutionalise his More Than Just Rare collection.

WORKING RELATIONS

During the 1950s, 1960s and 1970s, local suppliers to Constantin Wild would always come to the company offices on Fridays, delivering the roughs they had received themselves or the gems they had managed to cut during the week. "We were always fair to our suppliers, paying them the price they asked. We might have passed on certain stones or specimens, but we would never argue a set price down to the disadvantage of our supplier. That ensured that we always had first sight on the best new proposals, and I am proud to have kept up that tradition until today," Constantin Wild recalls.

TOURISTS BUYING ABROAD

At one point in time, Constantin Wild was receiving many visits from Germans who had bought parcels of very low-quality sapphires while abroad. Plied with promises of how the parcels could sell for double the amount the tourists paid for them, they had bought them in the hope of making a fortune back home. "Of course, the stones were of such poor quality that I never bought any – and people left highly offended when I told them that they had bought something worthless. Their insulted air taught me a lesson: later on, when victims of fraud offered me 'quality stones' like these, I would answer 'No thank you, I have enough stones as it is.' And the recommendation remains: buy only from reputable dealers you can trust. Many unconscionable individuals take advantage of the fact that you will never return to the place where you bought the stones to try and get a refund," he says.

GREAT STONES, GREAT STORIES

Major gem sales have marked important milestones in Constantin Wild's personal life and the company history – some can be told, others not, for various reasons. There were the demantoid garnets that got Constantin Wild out of debt after he had taken over the company from his mother, paying off his brothers at the same time; a spectacular aquamarine set paid for his divorce settlement; and a 75-carat imperial topaz sold to a Chinese client prompted him to start the tradition of the Gemstone Portraits and the More Than Just Rare collection. "About twenty years ago, I was exhibiting at the Tucson gem show, and I had brought this outstanding 80-carat cushion-cut Mexican red opal with me to put on display. Along comes this hillbilly, dressed in a plaid shirt with cowboy boots, and asks me casually over the counter: 'Hey, how much is that?' while pointing at the stone. As he didn't appear to be a serious customer, I just threw back, equally casually, '100,000 dollars'. He immediately replied: 'Make it 90,000 and I'll take it'. And he ended up buying it! I learned my lesson: don't judge a book by its cover - because he may end up snapping up your finest red Mexican fire opal," says Constantin Wild with a belly laugh.

One year, Constantin Wild had a large number of orders for green beryls from one customer - so many that he ran out of roughs and had to resort to what was on the market at the time. The market was in somewhat short supply, but: "Coincidentally, a dealer approached me at a show, offering me a rough Ukrainian green beryl weighing almost 1 kg. It would cost me a fortune and my margin would be very small, but I decided to buy it, as I knew I had a customer on hand. When I got the rough home, I looked at it in my workshop in Idar-Oberstein, and I suddenly noticed just what a beautiful crystal it actually was, with beautiful symmetrical edged surfaces. I fell in love with it so deeply then that I decided to keep it as it was - and to this day, I have not been able to part with it or let it be cut. In 2017, that particular beryl was highlighted as the 'object of the month' at the German Gemstone Museum, and it got a special exhibition all of its own," says Constantin Wild, fondly cradling the specimen in his hands.

The first time Constantin Wild travelled to mainland China was in 2010; he visited Beijing, Shanghai and Shenzhen to meet some of the largest manufacturers of jewellery, department stores and wholesalers. "My customers at the time were interested in large stones, and I ended up selling many of the major stones that I had collected over the course of many years. The Chinese clients were hard negotiators, and I have to admit that I made a big mistake, selling some outstanding gems below their actual value. One example was the 75-carat octagonal-cut multi-coloured imperial topaz mentioned above. However, the sale boosted my name in China, and it gave me the cash flow I needed to develop my company further. All the money was reinvested in the company, but I still regret letting that stone go to this very day. The stone was amazing for its colour, its size and its outstanding clarity. I had bought it for my own collection a few years before from a Brazilian dealer and had it re-cut for perfection. I tried to console myself by remembering that I had at least got to enjoy looking at it for a few years, but the sale of it prompted the reflection that so many great stones had gone through my hands – and I had no evidence of them. I had a great photograph of that stone, and from then on, I decided to have all my important stones immortalised with images and descriptions in my Gemstone Portraits and my More Than Just Rare collection," Constantin Wild explains.

BICOLOUR TOPAZ

75 carat

clockwise:
Exhibition poster at Beijing Empire Palace 'The Treasure Gallery'. Sunrise over Pudong, Shanghai. View inside the Forbidden City, Empire Palace Beijing. The 'Nest' Olympic Stadium in Beijing. Beijing at night looking towards Jingguang Bridge. Tina Hsu and Constantin Wild in front of the Forbidden City, Beijing 2011.

PIONEERS OF THE GEMSTONE BUSINESS AND SCIENCES

In 1932, Constantin Wild's uncle, a scientist and gem dealer by the name of George O. Wild (1894-1975), founded the German Gemmological Association, of which Constantin Wild is vice-president today. The association has developed into one of the internationally most renowned institutions of technical and scientific gemmology. "I was the first gemstone dealer to establish a virtual community for gemstone lovers; this comprised a web shop, a digital chat forum and a virtual meeting place for gemstone enthusiasts – I established Forum Gemstone back in 1998. I also served on the board of the International Colored Gemstone Association from 1999 to 2008. I was in charge of the gemstone information on the association's website, producing all the content for it, and that was at a time when companies and institutions didn't have those as standard," says Constantin Wild.

In Constantin Wild's life there is really no such thing as a normal work day. He travels for unusual gemstones, buying both rough and cut gems – the latter will most likely be re-cut in the family workshop. When working from his office, he oversees the cutting and selling transactions, liaising with the highest echelons of the international jewellery world, and in addition to exhibiting at fairs around the world, he also assists private and institutional collectors with consultations – finding and/or cutting particular gems for them. "Building business relationships naturally also takes place at social functions. When I attended an event in London, at a jewellery boutique, I struck up a long conversation with this very nice British gentleman about gemstones and the market for them in general. Later, it turned out that the gentleman was Prince Charles's private secretary," says Constantin Wild.

W. Constantin Wild & Co. has always been a globally active company, long before globalisation became a buzzword: always with an international orientation, active in several markets, simply because the gems were found in many different countries, and attracting the attention of many audiences. "Now I am responsible for the family gemstone venture, representing the tenth generation of my family. In 1847, my great-grandfather Johann Carl Wild IV, also known as 'Russ-Carl', founded the manufacturing facility that I have the honour of running today. The unbroken tradition of over 400 years is proof: I have the most wonderful profession in the world," he finishes.

"I have the most wonderful profession in the world."

Constantin Wild

GLO
BAL
ESTI
MA
TES

Value and rarity – a match made in gem heaven

by Nina Hald

No universally recognised grading system for coloured gems exists, and since supply and yearly production rates are impossible to estimate, their correlations with value are – to say the least – figures on which it is very difficult to comment.

"One consistent trend among gemstones is that the finer the quality, the greater the price appreciation over time. But only fine gemstones increase in value. In the top category – for the exceptionally rare gems – scarcity ensures that demand always outpaces supply," as Richard R. Drucker, publisher of the GemGuide, has written. The rarest collector class of gems has a tendency to escape all attempts at price-setting. "Such gems are so rarely encountered in the market that making a reliable price-quality index is not possible. Some of the gems known in the collection of Constantin Wild are of this category. These are one-of-a-kind gemstones and they just do not fit into standard grading scales," he writes.

Gary Roskin, director of the International Colored Gemstone Association, has written about the uses and appreciation of coloured gems, as well as their beauty, sizes and rarity. "It is a well documented fact that the quantity of diamonds mined each year exceeds the quantity of all the coloured gemstones put together by millions. The very few commercial-scale mining deposits of coloured gems we do encounter that allow for moderate to large-scale mining may produce gems for a maximum of decades. Of the tens of thousands of small, artisanally mined deposits, most will have been exhausted within a decade, with production in hundreds or thousands of carats, and not in millions of carats the way most diamond mine production figures are listed," he writes. Realising that also means that we are indeed looking at something truly extraordinary when we behold the most beautiful gems.

Estimating rarity on a global scale

By Gary Roskin, GIA GG, FGA
CEO of the International Colored Gemstone Association

For centuries, coloured gems were for royal ceremonial dress. They were the featured gems in the crowns, sceptres, orbs, swords, tiaras, bracelets, brooches, necklaces, rings, and robes of kings, queens and members of the royal family. These magnificent large, beautiful and rare gems, most commonly set accented by surrounds of diamonds, included emeralds, rubies, sapphires, spinels, amethysts, aquamarines, garnets, topazes, zircons and tourmalines.

Over the years, with more discoveries of artisanally mined gem deposits and globally distributed wealth, gems like these have also been worn by 'local royalty'. Political and entertainment dynasties, popular entertainers, corporate hierarchy, successful entrepreneurs and those of a certain affluence can now afford the luxury of owning and wearing these stunning natural transparent minerals.

First and foremost, gemstones are coveted for their beauty. But when we say that a gem is 'beautiful', what exactly is it that makes us so enamoured? A gem's beauty can be described by its colour, its shape and cut (two different variables), its natural perfection and, of course, its size. Large and colourful gems of fine clarity and artful cutting and polishing is what all of us are drawn to when we see such fabulous natural works of art.

Let's talk colour

Highly saturated vivid primary colours, light and subtle pastels, combinations of colour hues, or even something in between, each is its own unique blend, further defined by its lightness or darkness, softness or intensity. A gem's colour is the feature that determines its power as an immediate eye-catcher.

Adding to what Mother Nature has provided is the craftsmanship of the gem cutter. Taking the rough, raw gem material from the ground then determining how best to use such a natural wonder is just one of the gem artist's tasks.

Big

We cannot talk about a gem's beauty or importance without speaking of its size. Simply stated, the larger the gem, the more rare. It is rather intuitive that it should take much more effort from Mother Nature to create a large important gem than a small one. But 'big' is not always 'big', and sometimes 'big' is too big! Wearable gem materials can be discovered in a variety of sizes, from absolutely tiny to large enough to be seen from across the room. This is a unique variable when looking for that special gem. Size that may be quite common for one type of gem may be unheard-of for another. And speaking of size, we often confuse ourselves. Size is frequently expressed in carat weight. Weight and size are relatable, but they are not the same. After all, the size of

the gem is a measured factor, often being expressed in millimetres. The weight of the stone is commonly referred to in carats (1 carat = 0.2 grams). (The size of coloured gems depends on their specific density, which in turn depends on the elements they are made up of and how densely they are packed.) A 1-carat tourmaline is quite different in size to a 1-carat amethyst or a 1-carat opal, which is certainly different in size to a 1-carat diamond, all due to their differing densities.

Rarity

It is often misunderstood why coloured gems are much more affordable, when we know that they are much rarer than diamonds. It is a well documented fact that the quantity of diamonds mined each year exceeds the quantity of all the coloured gemstones put together by millions. The fact is that most diamonds can be commercially mined from exceedingly large deposits, with mines continuing to produce for decades, even centuries, making their production very predictable.

At the opposite end of this spectrum are coloured gems, with very few commercial-scale mining deposits. Those few larger gemstone deposits we do encounter that allow for moderate to large-scale mining may produce gems for decades, with only a handful still producing after centuries of extraction. Of the tens of thousands of small, artisanally mined deposits, most will have been exhausted within a decade, with production in the hundreds or thousands of carats, and not in millions as most diamond mine production figures are listed.

We make the comparison with diamonds because we are more familiar with them, through the decades of marketing efforts by DeBeers. But once we realise that coloured gems (including coloured diamonds) are much rarer, we realise that we are indeed looking at something truly extraordinary.

More on the numbers

The actual numbers of how many carats of coloured gemstones are mined each year are almost impossible to acquire, since most coloured gemstones (i.e. an estimated 95-98 per cent) are artisanally mined. In comparison, most diamonds are recovered by large corporate mining producers.

The reason for this centres on the differences in how Mother Nature creates gemstones. For example, diamonds are created under enormous heat and pressure, from 100 miles beneath the surface of the earth, and then blasted up through the crust in violent volcanic eruptions. In contrast, most coloured gemstones are created in a more peaceful and very random environmental process, where magma veins, forcing their way through the earth's crust, cook the surrounding mineral deposits. When these small rivers of heat encounter an area filled with the right minerals, gems can grow. Because of the random nature of magma veins finding pockets of gem-making minerals, most coloured gemstone deposits are small, with variable qualities of gem material.

It is so much easier to talk about the production figures of diamonds, since the bulk of them come from commercially mined ancient volcanic pipes. You simply look up the mining company's published annual production figures and add them to annual recovery numbers from other diamond mining producers. For example, the Rio Tinto mining company, which produces metals such as copper, salt and iron ore as well as diamonds, is the parent company of the Argyle Diamond Mine. Argyle has produced over 800 million carats of diamonds since opening in 1983 and produced 14 million carats in 2018. Figures like these cannot be collected for coloured gemstones.

Typically, the published numbers for coloured gemstones are given for 'gemstone-set jewellery'. These figures do not reveal how many gems are being used, what they are, what they weigh or what quality we are evaluating. But not all coloured gems are mined by artisanal miners. There are a small number of mining companies that produce colour, such as emeralds in Zambia and rubies in Mozambique. For example, we know that Gemfields produced 1.1 million carats at the Montepuez ruby mine in the first six months of 2019.

However, it is simply not possible to contact everyone mining rubies in Mozambique, sapphires in Madagascar or opals in Australia, just to calculate the total of what Mother Nature has provided. On the contrary, because coloured gems are mostly discovered by small groups or individual families, there's no desire to tally them up for statistics. Instead, they are quickly gathered up and taken to the local broker in the town nearby, until finally they come to the cities where they will be cut and polished, set in jewellery and sent on to the retail jeweller. Unlike diamond buyers, who have the luxury of ordering fine quality diamonds as needed, the coloured gemstone merchant has to buy gems from the artisanal miners who unearthed them in a process that is more or less coincidental.

The natural mystery

The fact that a thing is rare doesn't always mean that it is beautiful or valuable. But when you do have something rare, beautiful and valuable, treasure it.

Mother Nature does not always make it easy to find gemstone deposits, nor does she let us know how many beautiful and important gems we will recover from the site once it has been discovered. What we do know is the larger the gem, the more saturated the colour. But will there ever be another 100-carat mandarin garnet? Are there more mandarin garnet deposits waiting to be discovered? And if so, what will we find? These are the mysteries of Mother Nature.

right:
Amethyst crystals on mother rock.

Value developments in the coloured gemstone market

By Richard R. Drucker GIA GG, FGA (Honorary)
Publisher of the GemGuide

Natural gemstones are found in many qualities. Those that are sufficiently attractive to enter the gem trade are quality-ranked from commercial to extra fine. There is no universally recognised grading system for coloured gems, so dealers use various terms to describe the overall quality of gems. Some gem dealers for example might use terms such as top, gemmy or gem quality, while others have their own internal rating system that might include letters such as A, AA etc. For purposes of simple language and explanation here, the terms commercial, good, fine and extra fine will be used to classify gemstone quality. These quality ratings have been used for nearly 40 years in the context of price and market trend research. It should be noted, how-ever, that no quality rating scale can possibly account for the rarest collector class of gems. Such gems are so rarely encountered in the market that making a reliable price-quality index is not possible. Some of the gems known in the collection of Constantin Wild are in this category. These are one-of-a-kind gemstones and they just do not fit into standard grading scales.

One consistent trend among gemstones is that the finer the quality, i.e. extra fine, the greater the price appreciation over time. For most gemstone varieties, commercial quality is easy to source, but extra fine is much rarer and much more difficult. This will especially be true for the gemstones discussed here. And not all gemstones increase in value. Some gems that have had good supply in common categories may have increased slightly, stayed even or perhaps decreased over the past few decades. Again, it is important to note that this does not apply to the top, exceptionally rare gems. In this category, scarcity ensures that demand always outpaces supply.

AQUAMARINE

32 carat

Aquamarine

Aquamarine is the second most popular beryl after emerald. Although there are many sources of aquamarine in the world, the best known one is in Brazil, at the famous Santa Maria de Itabira mine in the state of Minas Gerais. Aquamarines in a saturated blue colour were found here at the beginning of the 20th century. The specific mine closed almost a century ago, and there are no Santa Maria aquamarines found in the market today, unless in vintage or estate pieces. Today, Santa Maria is used as a quality adjective for stones of that pleasant blue colour, but they are not from this mine.

In aquamarines, well cut stones can be sourced in the lighter blues, keeping those prices stable for most sizes. In the darker stones in fine to extra fine quality, and particularly in larger sizes, supply is good but not so plentiful as to keep prices down. As for the majority of aquamarines in the market, prices have risen over the last three decades, though less in recent years. Some categories have more than doubled in price. As for the prices of the Santa-Maria-coloured gems, with limited supply and superior quality characteristics, they have probably more than tripled over time. Santa-Maria-like colour has also been seen in some aquamarines from East Africa. These gemstones also receive a premium in the market.

DEMANTOID GARNET

3 carat

DEMANTOID GARNET

While there are a few sources of demantoid garnet, the most famous historical one is Russia. These gems possess a classic horsetail inclusion and ironically, the presence of these characteristic inclusions actually increases the value, whereas in most cases with gems, inclusions will decrease the value. These wispy inclusions of rutile look like a horse's tail, hence the description. Demantoid is the most expensive of the garnet group of gems. Its rich chromium-coloured green rivals that of the finest emeralds, but demantoids are far rarer.

Pricing demantoid over the past few decades has been challenging due to the rarity of the prized horsetail inclusions. As these collector gems command premiums, too few of the exceptional gems can be followed for pricing. There has been some increase in demantoid prices over the years, but it has not been as significant as the increase that applies to the exceptionally rare chrome-green specimens from Russia.

MANDARIN GARNET

103 carat

MANDARIN GARNET

Mandarin is the trade name for orange spessartine garnet. Its discovery was first reported in Namibia in the early 1990s, and it was notable for the pure orange hue of the material. Although orange spessartine garnets are known from other locations, many have subtle brown colour components. More recent deposits of orange spessartine in Nigeria have produced large, beautiful gems that can have a very nice yellowish-orange nuance, but usually with brownish overtones. Prices for happy, yellowish-orangey nuances continue to command a premium; look for shades like an orange.

IMPERIAL TOPAZ

45 carat

IMPERIAL TOPAZ

In nearly forty years in the gem industry, the finest imperial topaz that I have ever seen was the one in the collection of Constantin Wild – a most outstanding example of a true imperial with a perfect golden yellow colour and a truly intense red tip, fashioned in a pear shape. Few imperial topazes achieve such fine separation of the red from the golden colour. Yellowish-golden, orange to pink, pinkish-red and red topazes are all called imperial. The price increase is less for the yellowish tone but higher in pink, pinkish-red and orange-red shades.

Prices for imperial topazes have increased dramatically during the past three decades. The strong red material is rarely seen in the market. Currently, the most active markets for top material are in Asia and Europe. Since 1990, fine-quality imperial topaz has increased approximately four-fold. In the extra fine quality, prices have increased approximately five-fold in the same period.

PARAIBA TOURMALINE

4 carat

PARAIBA TOURMALINE

Paraíba tourmaline first appeared in the gem trade in 1989. The name is inspired by the source location in north-eastern Brazil. The material is unusual due to traces of the element copper. The finer grades exhibit intense 'neon' or 'electric' blue and green hues not typically seen in tourmaline.

When these gems first appeared in the trade, dealers thought they were 'overpriced'. On the one hand, they had an incredible look, but on the other, these were the highest prices ever realised for tourmaline. For the first ten years, prices were difficult to track, as they sometimes seemed to rise significantly in only a very short period, for example between the start and the end of a major trade fair such as the Tucson Gem Show.

Today, wholesale per-carat prices for fine and extra fine paraíba tourmaline surpass those of the vast majority of gem varieties. Prices have been tracked more closely over the past two decades. With so few large paraíbas in the trade above 3 carats, price reporting has been limited to stones under 3 carats. In the fine and extra fine quality grades, price increases in just the past 20 years have been five- to eight-fold. In recent years larger stones have been tracked for pricing purposes, but, as regards the increases in the same period, it is really only possible to speculate about them. Some say they have been at least ten-fold. From their debut in the market, it is only possible to speculate about how much they have risen. For fine and extra fine, it is fairly certain that the rise can be expressed in hundreds of per cent.

There are no new gem-quality or decent-sized Brazilian paraíba tourmalines available in the market. However, the cuprian tourmalines from Mozambique are rapidly undergoing the same development as the classic paraíba tourmalines.

RED SPINEL

5 carat

RED SPINEL

For many years, spinel was perhaps the most under-appreciated gem in our industry. Thirty years ago, those who knew of the gem often thought of it as an inexpensive alternative to ruby, and with the plentiful use of synthetic spinel in jewellery, the natural product was often overlooked. Over time, as consumers and jewellers began to appreciate spinel for the individual beautiful gem that it was, demand increased and, with that, prices increased as well. The beauty of spinel is that so many colours are available, and mostly they are untreated by heat or other methods. Red in particular has enjoyed the greatest price increase. In the past three decades, gems in the 3 to 10-carat sizes in extra fine quality have increased in price four-fold.

THE MAKING OF COLOUR

No cutting – no colour.
It is almost as simple as that.

by Nina Hald

It is the cutter who brings the natural beauty to life – the cutter's skills and expertise make the colour, inherent in the gemstone, bloom. It is a time-consuming process demanding great calm and patience. Short cuts? Impossible. The cutting and polishing process has its own pace and requires the extra mile many times over.

Yet sometimes, after just a few crucial seconds – the glittering prize! Colour unfolds all its beauty and dances on the walls of the cutting room at Constantin Wild. When the gemstone reveals the whole of its potential, when it shows its true identity, it is the moment of truth and the ultimate reward for the cutter's effort.

Computer programmes such as Gem-CAD may aid and optimise production, but adding soul to a stone is impossible without traditional handcraft and the experienced eye of the artisan. From rough to cut and polished, the cutter transforms the gemstones, already wonders of Nature, enhancing them to their ultimate splendour. A genuine sculptor of light, the gem cutter unveils the mysterious, mystical and magical radiance of the faceted gem.

As geological windows on the workings of the Earth, gemstones instantly take our imagination to faraway places. Constantin Wild is never too tired to travel to South America, Africa or Sri Lanka, where the rarest gems are mined. He does not find future treasures in safes, but in mines and warehouse basements. In countless little boxes, he takes raw stones like 'pebbles' back to Idar-Oberstein, the home of a centuries-old cutting tradition. Under the hands of his highly qualified craftsmen – master cutters – they will be transformed into some of the most valuable gems, which make leading jewellery brands and designers around the world gasp in awe.

Cutting makes the most of gemstones' intense, fabulous colours. A lapidary's confident hands shave away matter to preserve the essence. Later, the designers' visions are realised by the skills of goldsmiths and gem setters but, as countless professionals from the industry will testify, the beauty of a breathtaking jewel begins with the stone.

"IT IS ALWAYS AN EXCITING CHALLENGE TO TRANSFORM A NATURAL CRYSTAL INTO A PERFECT GEM."

Constantin Wild

Facets of a lapidary

Warm shimmer. Sparkling stars. Vivacious facets. The aim is always the same: excellence. Constantin Wild's masters achieve it by turning the rough gemstones – dull little rods of indefinable colour – into works of art beyond tangible value. The lapidary breathes life into transparent gemstones; he creates the facets which make them sparkle and shine. The art of faceting has been known since the 15th century. Many cutters specialise in particular gemstones, familiarising themselves with the varieties' specific characteristics. The lapidary is the key. None of today's modern aids can replace his or her long-time experience and patience, an eagle eye, a steady hand and the passion for 'his' or 'her' stone: giving a three-centimetre aquamarine almost 800 precise facets demands all the skills the lapidary possesses.

Pair of aquamarines
144 carat

Digging deep – standing proud

The old term 'stone carver' describes the profession of the engraver. The Egyptians created the first figurative three-dimensional depictions, for example the famous scarabs. The Greeks continued the cultivation of glyptic art. In Europe in the Middle Ages, seals became the indisputable signatures of the nobility.

Engravers cut images out of – or into – gemstones. If the motif is a raised relief, the result is known as a cameo. Intaglio is the opposite – a motif cut into the stone, mainly seen on signet rings. "As a gemstone merchant who is offered many cut and uncut stones from all over the world, I admire the skills and dexterity of German craftsmen," concludes Constantin Wild.

Signet ring of Constantin Wild
with blue moonstone

"LOVE OF BEAUTY IS TASTE THE CREATION OF BEAUTY IS ART."

Ralph Waldo Emerson

Marking and separating the crystal.

THE PROCESS

First, the crystal is 'sliced' – cut up into minor pieces. Razor-thin, high-speed rotating metal discs made of steel or brass sintered with diamonds are used for this purpose so as to waste as little as possible of the rough material. Velocities of 2,000 to 3,000 rotations per minute are common. Depending on the hardness of the particular gemstone, different discs are chosen. Water or oil is used to cool the gemstone and the rotating disc during the slicing process.

The slicing is followed by careful preforming – known by the French word *ébaucher*. Preforming means giving the gemstone its approximate shape, the top face and the underside. This is done by Constantin Wild together with the master cutter, using heavy vertical carborundum wheels with differing grains. The stone is given its silhouette, a most delicate part of the process. A computerised cutting machine is used for the work on the girdle so that it is completely symmetrical.

The preforming decisions are high-risk. The wastage when cutting a coloured gemstone is already considerable. It may be be-tween 50 and 90 per cent, depending on the crystal habit of the specific gemstone variety, growth axis, optical properties and refractive indices. In the blink of an eye, a poor decision during the preforming process may reduce a potential 20-carat gem to a 10-carat one. In other words: preforming is all about due diligence. It requires a strong lamp, a good magnifying glass, an eagle eye, imagination, the ability to think in abstract terms and courage.

The cutter leaves nothing to chance: he or she examines the gemstone meticulously, time after time, to make sure nothing gets overlooked. How is the colour situated in the stone? In which direction will it shine most beautifully? Which cut will bring out the best in the gem? It is not until this moment that the true value of the gemstone is manifested, its shape and character determined.

After the preforming, the cutting process can go in two different directions. There's the traditional way, where the gemstones are cut by hand with preformed stones mounted on wooden sticks, held down on a rotating metal wheel. This is done on sintered brass or high-tech rubber wheels revolving at 1,400 to 1,800 rotations per minute. Modern automated faceting is done entirely on CNC machines. However cutting by hand with the help of the quadrant is the marriage of past-time traditions and modern methods. The final polishing on each facet is done by hand with discs made of copper, zinc or lead, or a mixture of them – at a relatively low speed, just 600 rotations per minute.

It goes without saying that at Constantin Wild, the most expensive stones follow the traditional process. Cutting by hand is the only way to 'go with the stone', to react to the particularities of the gem and bring its inherent beauty to life.

"TO LOVE
BEAUTY
IS TO SEE
LIGHT."
Victor Hugo

Preforming the gems.

BRINGING OUT THE FACETS

Every faceted gemstone is composed of four parts: the table, the crown, the girdle and the pavilion.

With the gemstone glued on to the wooden stick, the girdle – the 'belt' around the gemstone – is done first. Then, the table and crown follow, and finally the pavilion. To determine a symmetrical scheme for the crown facets, the stone gets divided into sections. Depending on its cut, it may have six, eight or more sections, which are symmetrically mirrored over the crown.

During the faceting of the crown, the matching pavilion is made – which is why the gemstone has to be moved from the original dop, or wooden stick, to be held from the top. It is essential that the sections from the crown meet the sections from the pavil-ion, so that the facets line up from pavilion to crown (and vice versa). Not even marginal errors are tolerated.

Then the pavilion is faceted.

When faceting is completed, the polishing begins – both are done on horizontal laps made of copper, zinc or lead, revolving at 600 rotations per minute. The cutter must find every angle of all the many facets given to the stone and polish them one by one, with no interrupting or un-sharpening of the gemstone's symmetry, a skill that requires extensive training. After the pavilion has been polished, the stone is once again reversed, so that the crown facets and table can be polished as well. Table polishing is very delicate, as these are usually the biggest facets, which must be completely even. No marks from the polishing process must be left on the stone. When light shines on the stone, it should look totally flat, like a mirror.

The larger the stone, the more facets are needed to bring out its brilliancy and ensure the optimal colour and play of light.

For average gemstones, the entire process may take only a few hours. For the largest and most expensive gemstones, however, the cutting can easily go on for weeks. Before even laying a hand on the most precious stones, the cutters experiment with various options on copies of the same gemstone variety in less expensive material – it takes time and effort to define the ultimate splendour worthy of a kingly stone.

For the most important gemstones, Constantin Wild stays at the cutter's side throughout the entire cutting process, continuously examining and evaluating the progress. The process is stopped every other minute to verify the results of each single step.

For his company, Constantin Wild has developed its own ideal proportions and graduation of the curves of the bending facets on the stones. Ovals may have many different dimensions and shapes – as do cushion cuts. A collection of shapes appropriate for ovals and cushions is kept at the company because each gemstone variety requires its own particular modification of cutting patterns.

"Finally, comes this special moment, the climax, when the cutter hands me the cut gemstone with great pride. My evaluation begins; I double-check their work to approve the quality. Only then does it become part of my treasure trove," he says.

Very few select cut and polished gemstones actually make it that far. On any given day, there may be a million cut gemstones in the house of Constantin Wild, 2,000 of which would be the pride of any collector's possession or museum's display cases. The company's More Than Just Rare collection only comprises about 300 gemstones – almost impossible to buy from Constantin Wild. One needs good arguments to win him over and persuade him to let go. Money? Can't buy it!

"ELEGANCE
IS THE
ONLY BEAUTY
THAT NEVER
FADES."
Audrey Hepburn

Final pair of cut green beryl briolettes.

Stanislav Drokin

Round shapes, soft shimmer

The stone cutter is responsible for creating smooth surfaces, either flat or spherical, known as cabochons. This particular profession probably originated in India, world-famous for its rich gemstone deposits and traditions.

The smooth polished surface of a cabochon cut gives life and light to the beauty of inclusions. It brings out the shine in gemstones, which are usually either translucent or opaque. The skill of the cutter is essential; it defines the adularising, the shimmer of the moonstone, as well as the sparkling asterism of the star sapphire and the cat's eye effect (chatoyancy) in the chrysoberyl. And also, the colourful, radiant fire opals with their vivacious surface shimmer would be nothing without the talented hands of a dedicated stone cutter.

Moonstone

Ruby

Sapphire

Pleochroism – the case of several colours

Some gemstone types will display several colours – at least two, sometimes three – when viewed from different angles. This refers to different orientation axes in the crystalline structures of the gemstones. Tanzanite with its blue and purple axes and kunzite with its colourless and pink ones provide magnificent proof of this phenomenon. Tourmalines, rubies and beryls also display pleochroism, but it is less pronounced than that of the tanzanite and the kunzite. And as always, the final colour impression of the gem lies in the hands of the cutter, who must align the gemstone at exactly the right angle to bring out just the right colour play.

Tanzanite

Imperial topaz

Tourmaline

The agate cutters

The discovery of agates and jasper in the hilly Hunsrück at the end of the 15th century marked the beginning of the gem cutting tradition of Idar-Oberstein and the town's first cutting specialities. The cutting of onyx and jasper, along with malachite, turquoise and lapis lazuli followed. Agate cutting, as it is known, is a stone cutting speciality in its own right which differs greatly from lapidary work. When finished, the stones are either flat or have a slightly rounded surface. The opaque stones are used for rings and often seen in gentlemen's jewellery.

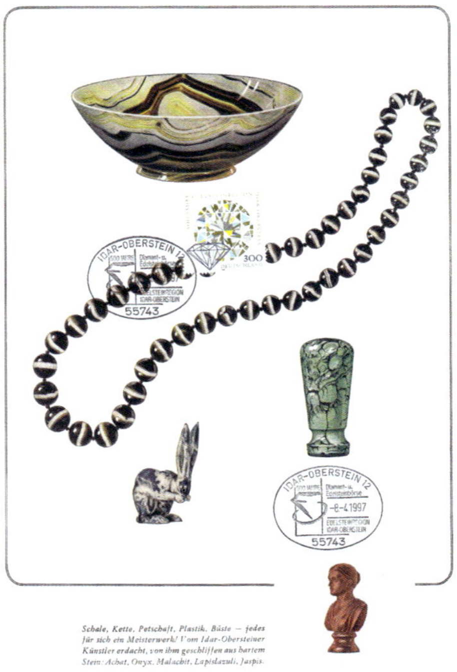

Agate bowl, necklace, animal carving, signet stamp and objets d'art, c.1949.

Art work

A multitude of *Kunstkammer* objects – bowls, chess sets, art objects, even currency and carved scarabs – have seen the light of day in Idar-Oberstein. Here, in Germany's world-famous gem cutting centre, some of the world's most spectacular coloured gems have been cut with exceptional virtuosity to the delight of their owners and wearers for the past 600 years. And to this day, you can still hear the whispers of Idar-Oberstein's stone cutters: their work lives for eternity. Frozen in perfection, conjuring the past, promising everlasting love and exuding the luxury that Nature designed.

Kingfisher sculpture carved from lapis lazuli and white and orange opal with onyx eyes and obsidian beak mounted on tourmaline and obsidian base with 18k gold.

Cameo carving of a lady in black and white sardonyx.

JOY
OF
COL
OUR

JOY OF COLOUR

by Nina Hald

Colours are the epitome of timeless beauty and have been the summation of gemstone nirvana for millennia.

As the most valuable creations of Nature, gemstones of transcendental colours have found their way into our history of culture and art. The topic of gemstones entails theories of their origins (geology, physics and chemistry), tales of traditions (religion and history) and stories of adventures and aesthetics (art and culture) – but pure appreciation of gemstones is always related to their joyful colours.

Finding two coloured gems that match is an achievement because, unlike diamonds, no two gems are identical. In itself, matching a pair is quite a feat. Now, imagine the work and time that go into finding two dozen matching gems for the composition of a necklace ...!

A perfectly matched suite of gems on a velvet tray in a black box looks as if it has been put together effortlessly. However, the reality of all the effort that goes into the compilation of that set is unseen and most likely unknown.

THE ARTIST'S EYE

First, the assembly of a gemstone suite demands an extensive inventory. Basically, there must be an abundance of gems to choose from. Sometimes it takes years to collect the gems needed for a suite. Secondly, there must be a vision, a profound idea, behind the 'layout' – another designation for such a collection of stones. No vision – no wow factor. In other words, there must be a guiding principle that appeals to the eye of the beholder. Order and system bestow balance on beauty. Each individual gem may very well be gorgeous in itself, but that 'does not a suite make'. Thirdly, it takes an artist's eye to make the final matches, especially when it comes to prestige layouts composed of many different gemstone varieties, colours and cuts. However, we are getting ahead of ourselves. Let's begin at the beginning – how to make a match.

Some layouts are done with just one gemstone type – for example tanzanites. Layouts may be made according to size; either each gem of the same size or graduated in size. Other layouts are done in the same colour – some with graduations in colours – some even in contrasting colours.

Pairs are usually presented with a view to being set as earrings. At other times, a pair is an initial basis for building a larger collection. Extraordinary pairs usually sell like hot cakes before they can be augmented with more. A masterful example was a sky-blue set – "like day and night", as Constantin Wild described the pair, comprised of a blue-violet emerald-cut 43-carat tanzanite and an azure blue emerald-cut 38-carat aquamarine. The two gems interacted superbly, catching each other's colour flashes in the optimum way.

A suite can be composed of as 'few' as three gemstones – a pair for earrings and a pendant stone, which of course can also be set in a ring. The largest suites are arranged with a selection of gems for a necklace, a smaller group set aside for earrings, one gem intended for a ring (or a pendant) and often an additional selection of gems for a bracelet. If the intention is to make up a parure, more matches must be achieved.

"On average, we assemble six to eight outstanding layouts every year with various concepts. Some with mixed colours across gemstone varieties but with the same cut, others with mixed shapes that still go well together. All may well have the same size and cut but in different colours and gems that complement one another in happy springtime colours, like pink tourmaline, peridot, mandarin garnet, chrysoberyl and kunzite, others in autumn colours like yellow-brownish nuances," says Constantin Wild. However, he is also known to be a fan of 'blue moods' composed of tanzanites, aquamarines, blue zircons and light and dark amethysts.

The layouts require all colours of the spectrum to be present under the same roof. "We get to be creative in ways that not many others are able to. When we put an unexpected colour match in front of some buyers, they must have it immediately, because they have never seen anything like it. It makes the stones even more unique. For that reason, I also refuse to break up some layouts if the buyer only wants one of the gems – because it would be a crime to split up a perfect match. This business and the particular art of assembling layouts are about love," says Constantin Wild, who has even given some of the most outstanding layouts names.

PAIR OF TANZANITE AND
AQUAMARINE IN OCTAGON CUT.

43 and 38 carat

A 'normal' layout for Constantin Wild will be made up of anything from 10 to 24 gemstones, intended for a set comprised of a ring, a necklace and a pair of earrings. A layout of a necklace with a frontal decoration will usually be made up of an uneven number of gemstones.

Once, as a challenge, Constantin Wild put together a prestigious layout of cushion-cut gems: a blue-blue aquamarine, a pastel-green tourmaline, a golden beryl, an orange mandarin garnet, a grass-green peridot, a yellowish-green chrysoberyl, a pinkish-red rubellite, a purple cuprian tourmaline, a lemon-yellow canary tourmaline, a vibrant yellowish-green sphene, a saturated pink kunzite and a purplish-blue tanzanite. It took years to put together because the intensities of colours and sizes of the gemstones all had to match. To realise something as ambitious as this, gemstones that have already been cut may sometimes be re-cut to match the other gems in a layout; at other times, new individual roughs will be cut specifically to complete the layout.

For some of the most outstanding gems, Constantin Wild may sometimes ask a jewellery artist to draw suggestions as to how they could be mounted.

Like most jewellery companies that have developed their signature colour combinations, Constantin Wild has a design guide for the composition of layouts. "I like to give designers ideas or some inspiration as to how gemstones can be combined, but I never go overboard. I am never too specific. Because in the end, it is up to a designer to arrange the layouts in his or her own style. I never presume to design the final look. I always leave melee stones and side stones out of my layouts, in order to offer only the centre stones," he says, explaining his way of showing respect for the designers' work.

"When I decorate my showroom or my booth at a trade fair with single gemstones, I usually carry the gems each in a separate box, grouped on trays of six. Even at this point, I am trying to indicate six stones that somehow match or correspond with one another, either in size, colour or variety. At this stage, they are already suggesting how they could be matched with others," he explains.

"I always leave melee stones and side stones out of my layouts, in order to offer only the centre stones."

Constantin Wild

SUITE OF 28 AQUAMARINES FOR
NECKLACE, EARRINGS AND RING.

57 carat

THE BEST PUBLICIST

Constantin Wild recognises that diamonds have always had the 'better publicist'. "In a way," he says, "diamonds are the roses of the gemstone world. But thankfully, Nature has given us many more flowers than just roses – flowers more beautiful, more colourful and much rarer than roses." Therefore, he sees it very much as a question of enlightening people as to the existing huge variety of choices besides 'white roses'. "I see diamonds being marketed and sold as the hardest stone in the world – but they are certainly not the rarest," Constantin Wild says. Coloured gems of excellent quality are rare to the point of being as rare as hens' teeth. Meeting demand is usually impossible. "I always advise people to seize coloured gems when they are on offer. Gems show colours in such saturations that they almost make your heart stop beating. In comparison, the colour in so-called coloured diamonds is sometimes so delicate that you need a certificate to tell you what colour they are supposed to be. With coloured gems, there is no doubt as to their blue, pink, green, red, orange or purple shades," says Constantin Wild with great finality.

'AUTUMN COLOURS SUITE' ARRANGEMENT
COMPOSED OF 18 OFF-SHADE CITRINES AND QUARTZ.

142 carat

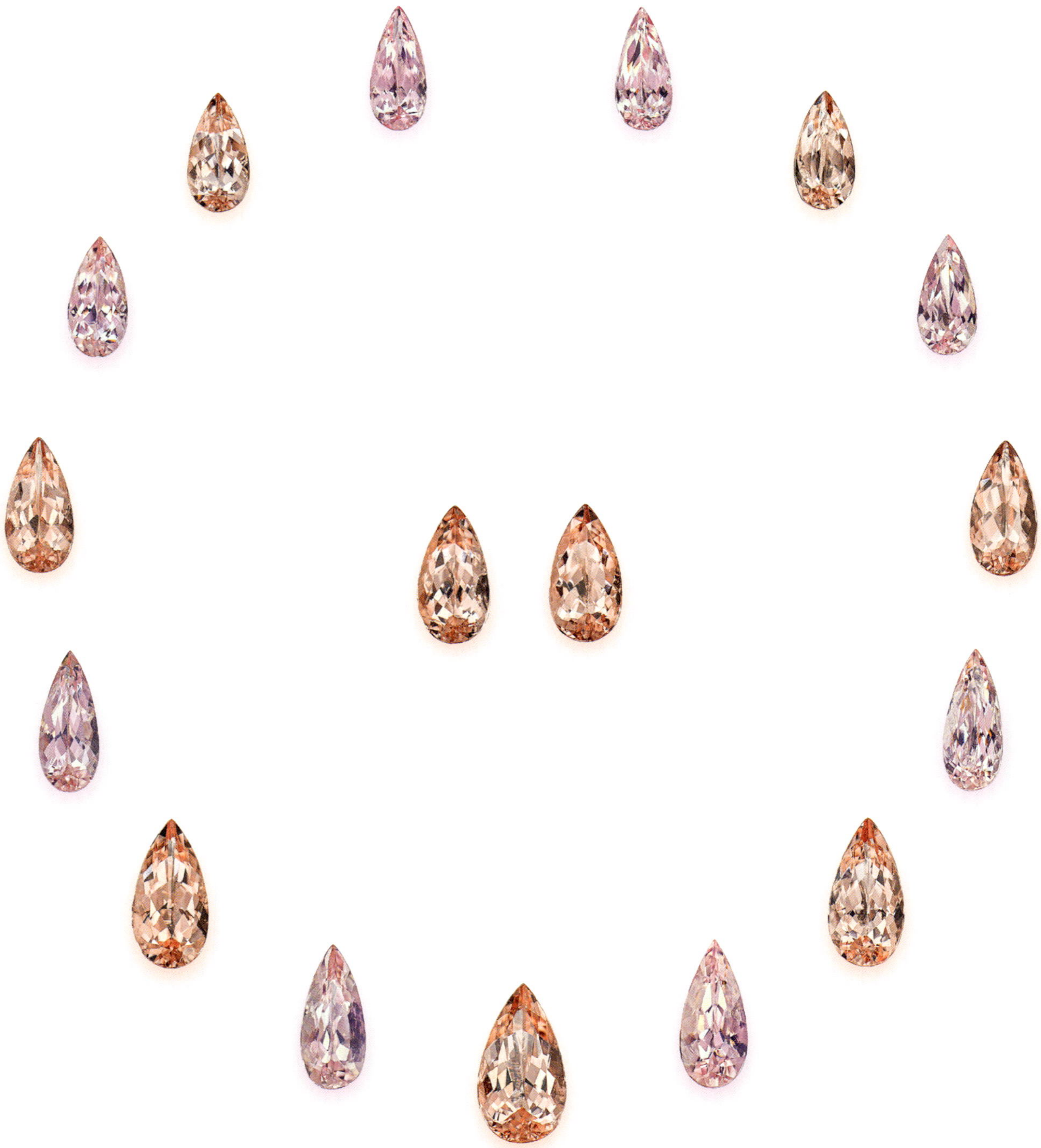

'LA VIE EN ROSE' SUITE COMPOSED OF
17 PEACH- AND ROSE-COLOURED MORGANITES.

97 carat

AQUAMARINE SUITE COMPOSED OF
15 PEAR-SHAPED CUT AQUAMARINES.

51 carat

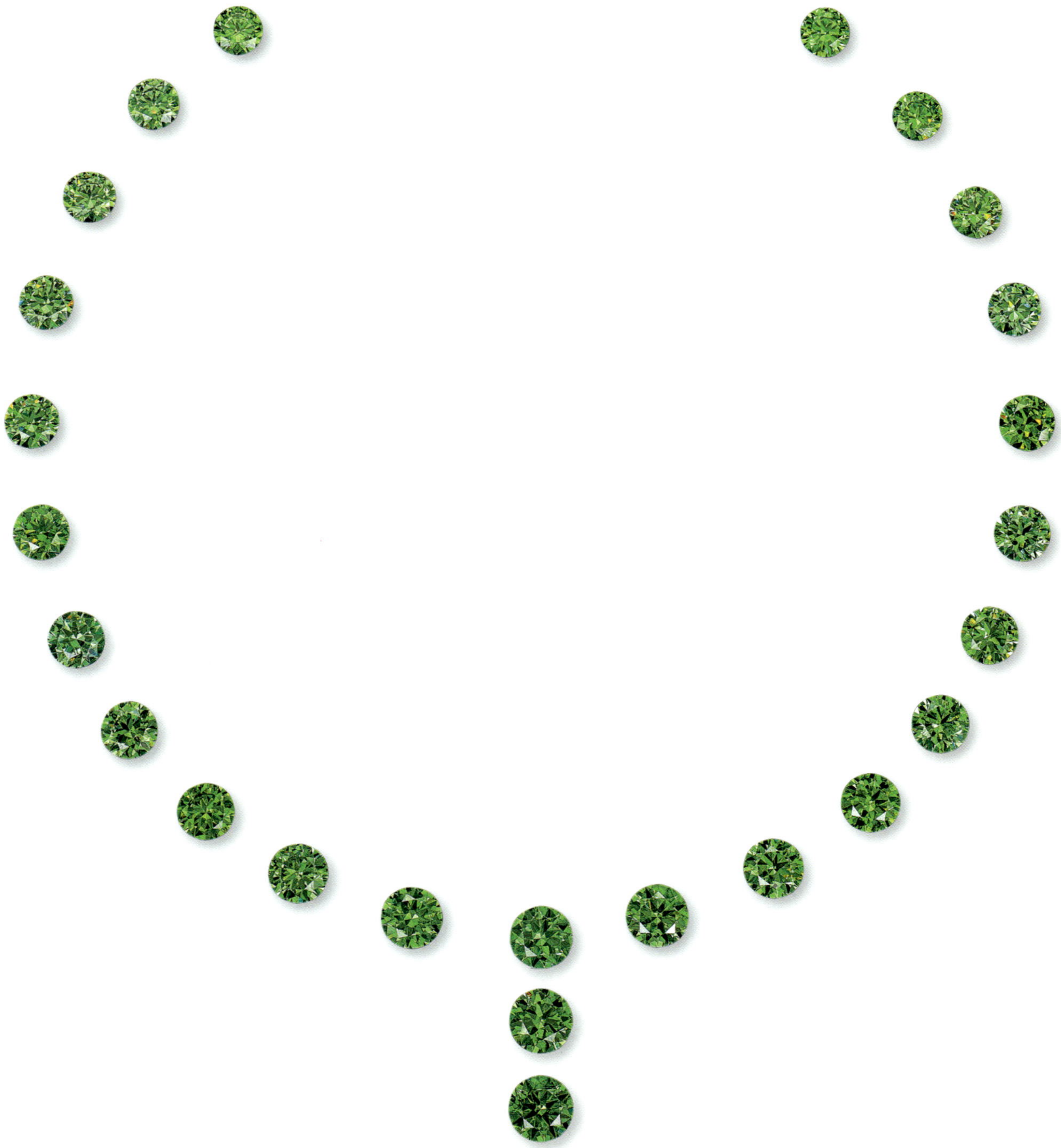

GREEN GARNET SUITE COMPOSED OF
25 RUSSIAN CHROME DEMANTOIDS.

40 carat

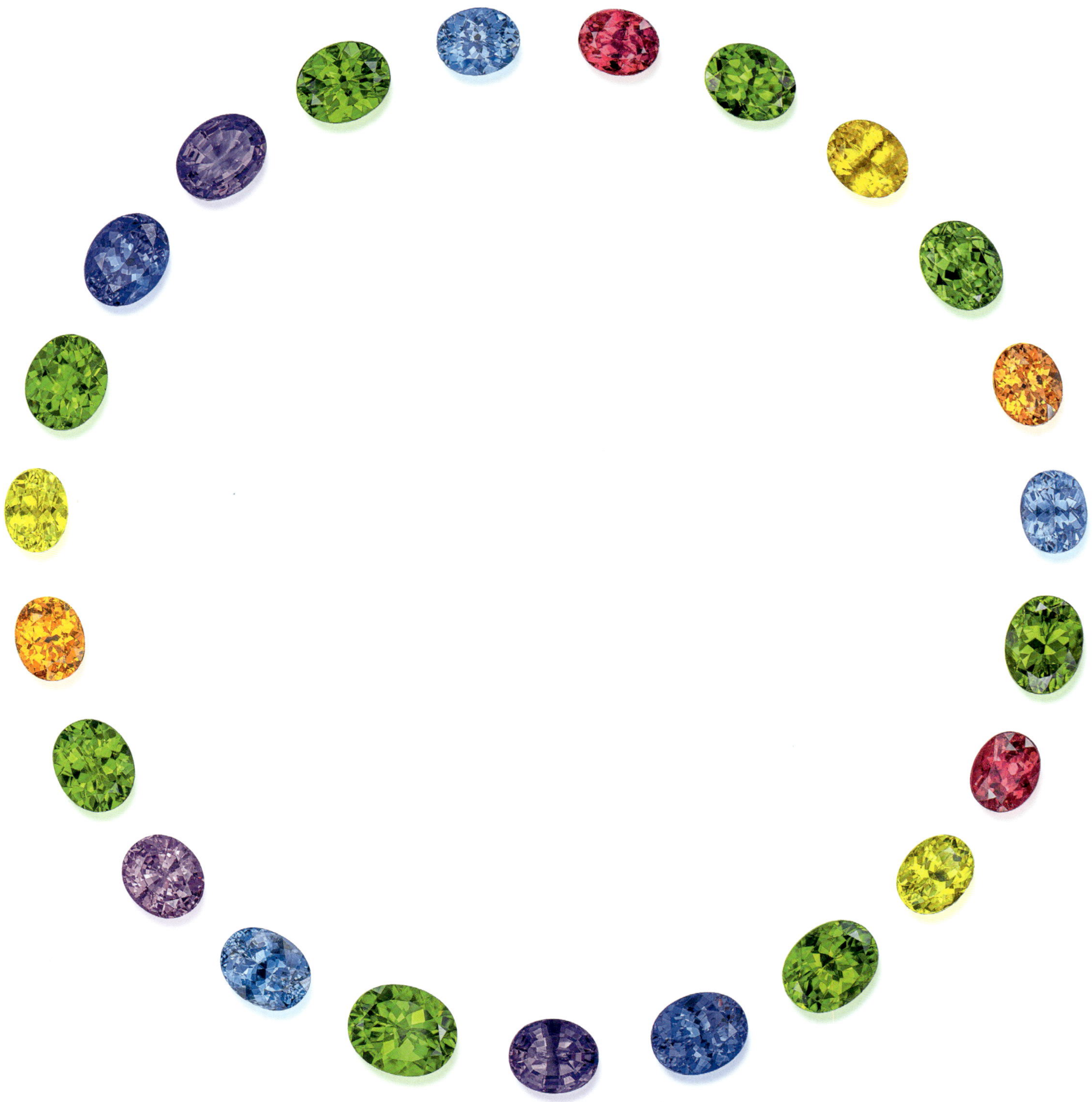

'JOY OF COLOUR' SUITE, 23 GEMS ARRANGED FOR A COLOURFUL NECKLACE COMPOSED OF RUBELLITE, PERIDOT, GOLDEN BERYL, MANDARIN GARNET, AQUAMARINE, CANARY TOURMALINE, TANZANITE, AMETHYST AND PURPLE SAPPHIRE.

110 carat

'BLUE MOOD SUITE' 18-GEM ARRANGEMENT COMPOSED OF TANZANITES, AQUAMARINES, KUNZITES AND BLUE ZIRCONS.

154 carat

'RHAPSODY IN BLUE', 17 BLUE GEMS ARRANGED FOR A PASTEL AND INTENSELY COLOURED NECKLACE COMPOSED OF TANZANITES AND AQUAMARINES.

145 carat

'PASTEL STONEHENGE'. 20 GEMS ARRANGED FOR A PASTEL-COLOURED NECKLACE COMPOSED OF AQUAMARINE, IMPERIAL TOPAZ, GREEN BERYL, MORGANITE, LILAC AMETHYST, GOLDEN BERYL AND COGNAC BERYL.

110 carat

MULTI-COLOURED 16-GEM ARRANGEMENT COMPOSED OF RUBELLITE, AMETHYST, KUNZITE, TANZANITE, GOLDEN BERYL, BLUE ZIRCON, GREEN TOURMALINE, CHRYSOBERYL, MANDARIN GARNET AND AQUAMARINE.

300 carat

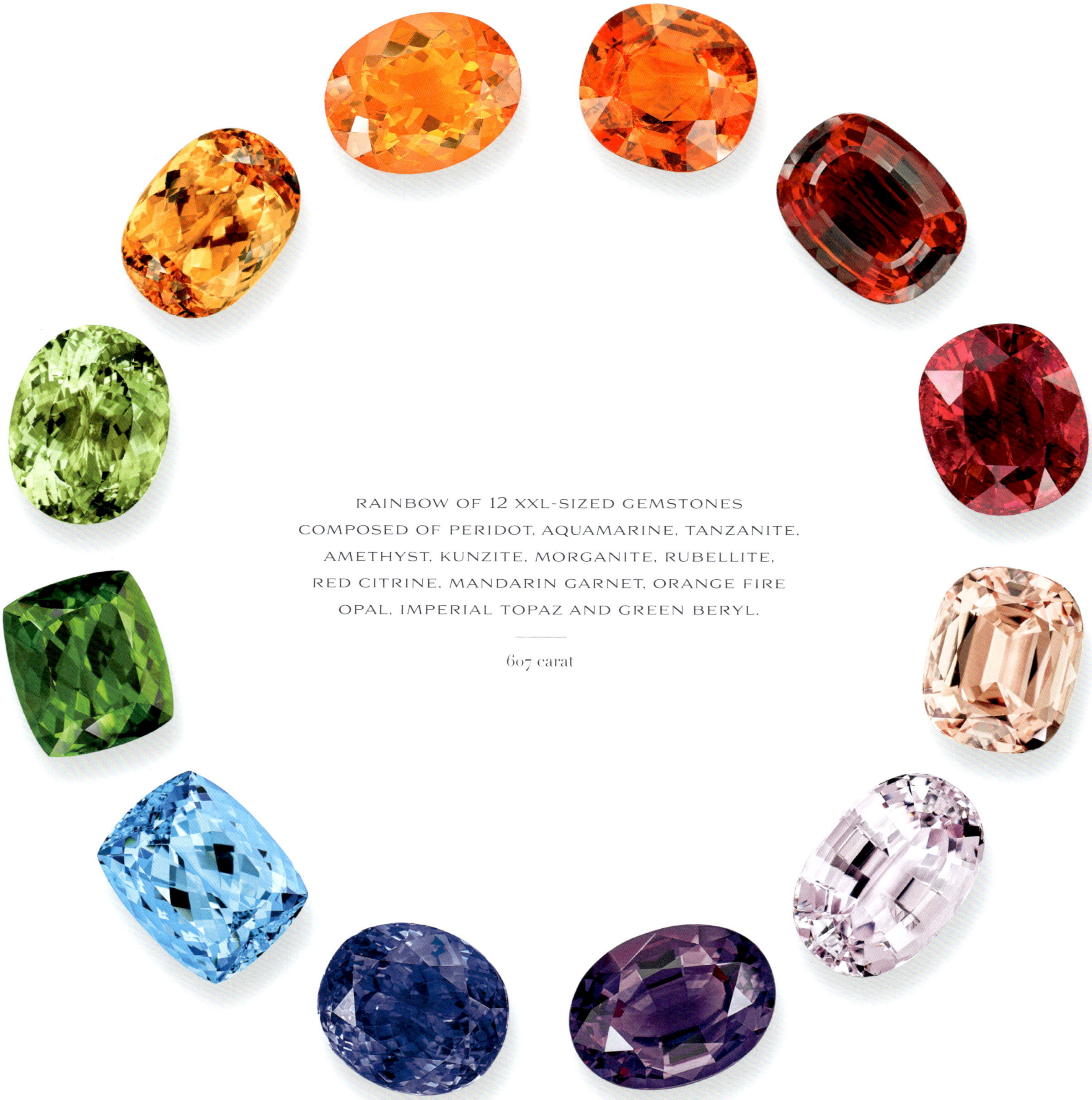

RAINBOW OF 12 XXL-SIZED GEMSTONES COMPOSED OF PERIDOT, AQUAMARINE, TANZANITE, AMETHYST, KUNZITE, MORGANITE, RUBELLITE, RED CITRINE, MANDARIN GARNET, ORANGE FIRE OPAL, IMPERIAL TOPAZ AND GREEN BERYL.

607 carat

A RAINBOW SELECTION OF 600 CARATS

A layout most certainly does not have to be composed of the same gemstone, in the same size and cut. In an expertly matched layout, a multitude of cuts and hues can be displayed. Constantin Wild demonstrated this in 2017, when he received a very special call in his career as a gemstone merchant. An important jewellery-making client wanted to dazzle with something opulent and expensive. "I went to work immediately. I pulled together some of the largest stones I had at the time. I put them on a table in front of me, contemplating how to combine them," Constantin Wild recalls. The twelve gemstones made up a total of 607 carats: a peridot from Pakistan, an aquamarine from Brazil, a super-blue tanzanite, a vivid purple amethyst from Brazil, a soft pink kunzite from Afghanistan, a pink morganite from Brazil, a pinkish-red rubellite from Mozambique, a top orange-red citrine from Brazil, a juicy orange mandarin garnet from Mozambique, a bright orange fire opal from Brazil, an orange imperial topaz from Brazil and a tourmaline-green beryl from Brazil. "The gemstones were all either oval or cushion, and when composed together in this rainbow selection, they turned out to work outstandingly well. It was a truly daring colour combination – and I have never seen anything like it during my time in the business," says Constantin Wild with justifiable pride.

COLOURFUL MELEE

Traditionally, coloured gems set as centre stones in high jewellery are highlighted and even optically enlarged by being surrounded by small, colourless diamonds. But there are also other ways to add panache and glamour to a creation – a trend that has been gaining ground for the past decade, especially in Asia. Cut multicolour gemstones in tiny sizes – also known as melee – offer a huge variety of colours while focusing attention on the colourful centre stones. "As the centre stones usually contain several colour flashes due to pleochroism, it can be a lot of fun to accentuate the various shades by complementing, combining and repeating the colour of the centre stone with a mix of matching melee. For example, a large kunzite as the centre stone can be highlighted with melee in pink sapphire, pink imperial topaz and pink spinel. A violet sapphire can be combined with melee in purple spinel, purple sapphire melee and/or amethyst melee, and even greyish melee can complement it quite beautifully," Constantin Wild explains, leafing through the large selection of melee that he keeps on file in many saturations for the purpose.

TIFFANY
SPRING COLLECTION

IMPERIAL TOPAZ
MULTI-COLOURED PEAR SHAPE

THE ONES THAT MUST NOT GET AWAY

"Sometimes at another gemstone dealer's I will see a stone that I know will complete one of my layouts. Then I simply must have it," says Constantin Wild with something akin to an obsessive look in his eyes. To obtain the stone in question, Wild dare not let on how much he wants it.

"When I see a gemstone that appeals to me, I have to touch it. Feel it. I do not necessarily want to look at it through the loupe; I like looking at gems with my eyes. I do not use my tweezers. I just let the gem roll playfully through my fingers. If it is the just-right stone, I feel a strong desire to own it. Maybe even a little nervous of losing the opportunity to acquire it. At that time, the seller is close to making the sale. I have to hide my feelings so that the seller does not notice that I have already fallen in love with the stone and want it for my collection. Otherwise I cannot negotiate a decent price. Usually, I borrow the stone for a few hours, to look at it under various lighting conditions and moods," Constantin Wild explains.

Just as when we fall in love, the first fraction of a second of looking at a gem is all-important – and usually Constantin Wild is right in his assumptions. "I am presented with gems in all price categories – from a few thousand to several million," he says. "I can borrow them to clarify my first impressions without signing for them, because everyone knows that I am trustworthy and serious about my business, and that I know what I am doing. I honour deals and reasonably set prices. That is why I am in the fortunate position of having the first sight when something special comes on to the market. However, I usually buy less than 0.1 per cent of the stones that are presented to me."

When Constantin Wild walks through markets and trade fairs, he scans all the gems displayed, and by eyesight only, he can pick out the few that he wants to acquire. "Most merchants will have 99 per cent inferior gems on their trays, and I am only interested in that really good one per cent," he says. He has developed his eye for gems over the past forty years in the business – in other words, this is a process that has been going on for most of his life.

"Generally, I have a good eye, an understanding of colours and the proportions of the gemstones," says Constantin Wild. "My eye for beauty is not limited to gemstones either, and this helps me in my profession. The people to whom I sell gemstones – the designers and creative goldsmiths – have the same ability. We communicate with great ease, even non-verbally, over the gemstones."

Once, Constantin Wild asked his father, Fritz, what his favourite gemstone was. Fritz replied that the imperial topaz was his favourite. "So, while I was in Brazil on my first business trip, in my early 20s, I checked the street market for imperial topazes. I very much liked an 8-10-carat imperial topaz, but money was tight for me at the time, and I only had a small amount to spend on total purchases. I called my mother, begging her to allow me to spend additional funds on that one orange imperial topaz. Luckily, I was able to persuade her of my love for the stone. I acquired it and sold it again very soon after. Since then, I have followed that rule of thumb. When I see a stone that I want to keep for myself, I know it will always appeal to someone else too. Sure enough, those are the gems that sell the fastest. In the case of the imperial topaz, it gave me confidence to follow my taste, eye and inner desire in the selection of stones on my subsequent sales trips," Constantin Wild recalls.

THE MERCHANT AS A MATCHMAKER

In the international gemstone and jewellery industry, it is a well known fact that Constantin Wild has a soft spot for imperial topazes. "Some imperial topazes display red flashes, others pink. So, when I am able to match two imperial topazes down to the very last nuance, it gives me a deep sense of harmony and pride. It is a combination of luck, coincidence and being in the right place at the right time. And on top of it all: to have Nature give us those two specimens, which may not necessarily have been found side by side but still match each other to perfection, that is a moment in my life as a merchant and lover of gemstones when I get to play the ultimate matchmaker," he says with a hint of deep satisfaction. When it comes to imperial topazes the sky is truly the limit for him. "I once collected fourteen imperial topazes over the course of twenty years; each had a slightly different shade – from orange to golden to pinkish-orange to pink and red. When displayed next to one another, they made up the most stunning set," he says. For him, it became paramount to showcase every possible nuance imperial topaz could display – and with those fourteen gems, he succeeded.

SET OF 14 IMPERIAL TOPAZES

198 carat

MARIE ANTOINETTE'S PINK TOPAZ

A gem with supposedly royal provenance found its way into the possession of Constantin Wild's family more than a generation ago: a deep pink topaz of some 18 carats that is said to have belonged to Queen Marie Antoinette (1755-1793). It is set in a ring that once graced the hand of Constantin Wild's mother. "My father told us the story about how he bought the pink topaz in Paris in his youth. At the time, our company had an office in Paris, and my father even lived there for quite a while. He bought the topaz before he and my mother knew each other," says Constantin Wild. His parents married in 1962. Fritz Wild had the gem re-mounted into a ring. "My father always stressed that I should take good care of this particular gem because of its provenance. Pink was my mother's favourite colour, and even though it was given as an engagement ring, she reserved it for fancy formal use."

Rohstoffquellen der Fa. W. Constantin

THE ONES THAT SHOULD HAVE GOT AWAY

In some instances, Constantin Wild has been fooled by colours – lessons learned, as he calls the episodes. If a gemstone merchant claims that this has never happened to him or her, that person is simply not trustworthy in Wild's opinion. Luckily, with both feet solidly planted on German ground, Constantin Wild acknowledges the few times he has been wrong, either lacking the necessary experience as a young man in the business or because he was cheated by con artists.

"I once bought a 10-carat orange sapphire in Australia. It cost me a fortune back then, but it was amazing, and I had to have it. When I looked at it in Idar-Oberstein light, it turned out to be a brownish-yellow stone! I was so disappointed. It had looked glorious in Australia, but not so in the light of the northern hemisphere. I felt bad spending so much money on what had turned out to be an inferior gem, and I ended up returning it," says Constantin Wild, sharing an incident from his youth.

Years later, more experienced and on home turf, he approached an opportunity to buy some opals with much more caution. It happened when Ethiopian opals had just come on to the market. A stranger came to Constantin Wild's offices in Idar-Oberstein. "At the reception desk he showed me a handful of these opal roughs: clear crystals with full plays of colour inside. They looked almost like children's toys. The outside of the opals had been oiled, to allow an insight into the stone. I told the stranger that to my knowledge Ethiopian opals were not stable and were at risk of breaking," he says. The stranger pretended not to understand him, and then to have no knowledge of what he was talking about. That in itself raised a certain level of suspicion. "In the end," he continues, "I bought one of the opals that afternoon. I looked at it the whole evening, contemplating what to cut out of it – perhaps a pair, a single gem, a cabochon or a free baroque organic shape."

"The next morning, when I came back to my desk, I was rather shocked to find the rough completely cracked, almost falling apart. It had obviously been pulled out of the liquid just before being shown to me."

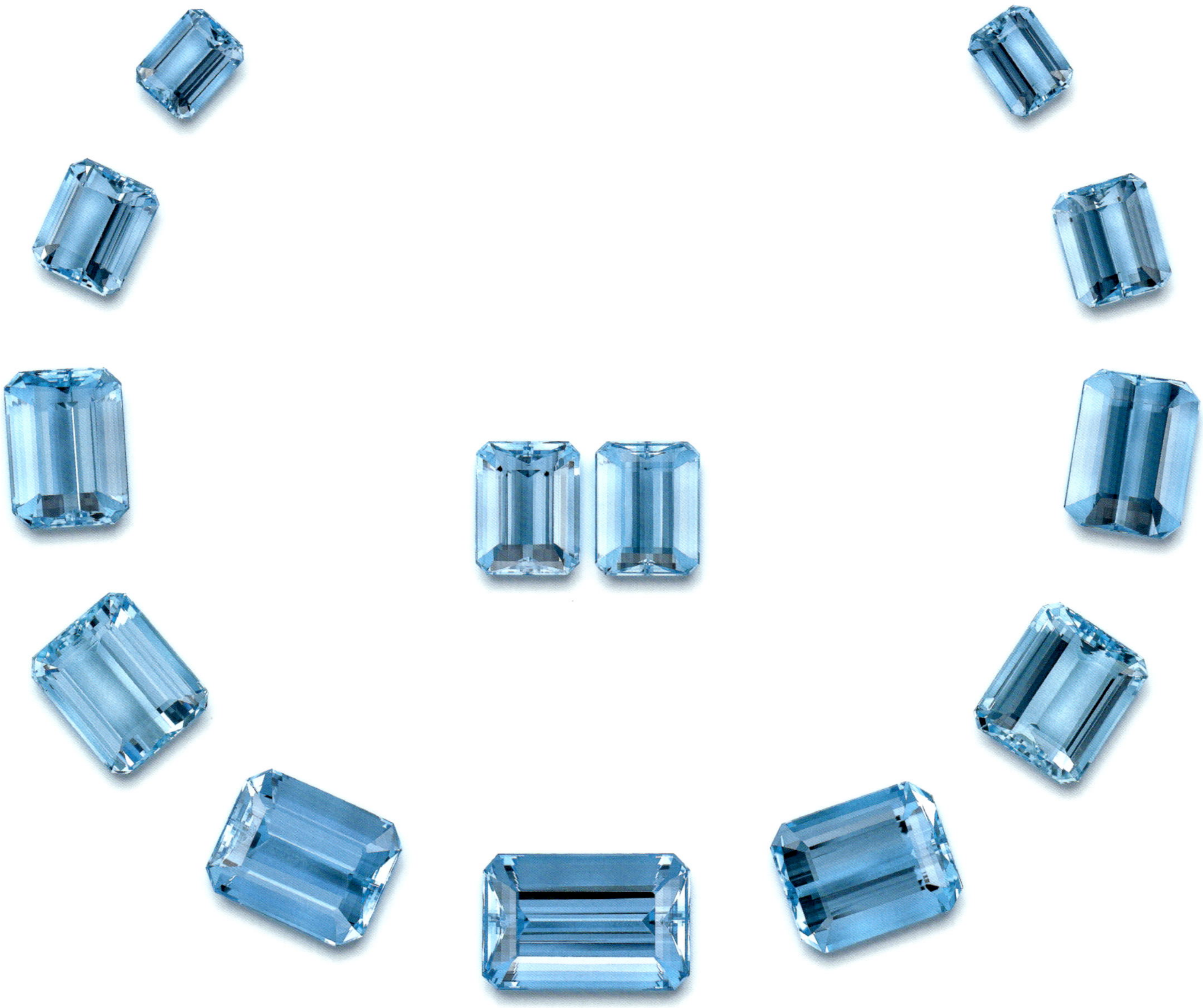

AQUAMARINE SET

380 carat

THE DIVORCE SETTLEMENT SET

With more than four decades in the business, Constantin Wild remembers many gemstones and the circumstances under which they were sold. Being in love with his treasures, he finds it can be hard to part with some of them, but knowing that they will be displayed in magnificent jewellery is always a consolation. "Sadly, at a certain time in my life, I was heading for divorce," he recalls. "When the time came to arrange the settlement, I had to raise some money. There was a spectacular set of aquamarines that I had had in my possession for some time. It had taken me several years to put the selection together, and it was superb in every way – colours, size, cut, the aquamarines were a match made in Heaven, although my marriage, unfortunately, had not been. An Italian jewellery designer – head of design for one of the big brands – expressed an interest in seeing the set, so I flew to Italy myself to present it to her. She of course loved the aquamarines very much and bought the layout immediately. The tidy sum from that deal was just enough to pay the settlement."

PORTRAIT OF A GEMSTONE

For the past ten years, Constantin Wild has collected exceptional gemstones selling from 10,000 dollars and upwards. Lab reports, images and background descriptions are gathered together in what he calls a Gemstone Portrait. "In the Gemstone Portraits, the gems are described by their physical properties with photographs as documentation, to which I add a brief account of what I find special about each one. This account is more comprehensive than what a normal lab report would or could document. It is my opinion about the stone. I began doing portraits on exceptional gems in 2011 and now have about 400," says Constantin Wild. This portrait gallery also began because he was sad to see some of the gemstones leave. "I had loved and appreciated them, so I wanted a memento of them. An unexpected benefit of this, of course, is that I can also verify a gem's reappearance on the market. That way, I am able to document that I have owned and sold it before."

Constantin Wild

GEMS · TRUST · WILD

GEMSTONE PORTRAIT

Reference Number 00762

IMPERIAL TOPAZ

one faceted gemstone
natural topaz
pear shape
star cut

20.26 carat
26.45 x 10.38 x 10.04 mm

brownish-pinkish, reddish-orange
Brazil
natural colour

DESCRIPTION

In 1735, some topazes found in the Ouro Preto region of Brazil were so beautiful that the epithet ‚imperial' seemed appropriate. Imperial topazes normally come in colours ranging from yellow to orange. However, the most sought-after specimens are stones with pinkish and reddish hues. This amazing imperial topaz features a mixture of colours in one stone: from a brownish-pink to the highly coveted reddish-orange. A one-of-a-kind beauty.

DOCUMENTS

Accompanied by DSEF GEMSTONE REPORT NO. 020995

Paloma Picasso for TIFFANY, Rings, Holiday Edition

JEWELLERY HISTORIAN VIVIENNE BECKER SHARES HER PERSPECTIVES ON HOW THE PAST YEARS OF NEW GEM DISCOVERIES HAVE CHANGED HIGH JEWELLERY DESIGN – AND HOW SELECT JEWELLERY DESIGNERS, BRANDS AND COMPANIES USE COLOURED GEMS.

by Nina Hald

"I believe the discoveries of new gem varieties such as tanzanite, tsavorite and malaia garnet have been among the most important happenings in recent jewellery history. They have been real gemmological and artistic turning points," says Vivienne Becker. In her opinion, most of the stones unearthed – in all their exquisite, previously undreamt-of colours, tones and nuances – have given designers the broadest colour palette ever.

"These intriguing stones have also broken down arbitrary classifications and outdated prejudices about 'preciousness'. All gemstones are rare, precious natural resources. The new stones have also broken all the rules of colour in jewellery – previously limited to emerald, ruby and blue sapphire – to enable jewellers to create imaginative, painterly effects, to be more expressive, to infuse jewels with emotion and story-telling," Vivienne Becker explains.

"The most recently discovered gemstones have allowed designs to be more tonal – tone on tone – with subtle shading, nuances and different textures of light as well as shades of colour. I believe the new gemstones have liberated creativity in jewellery design as never before – bringing freedom, fantasy and artistry. One of the wonderful things about coloured stones is that each designer – and indeed each jewellery client – responds to different colours in different ways. Colour is perhaps the most powerful emotional trigger in a jewel. Each designer brings his or her own emotions, responses, visions to a jewel, creating an intensely individual interpretation of colour," says Vivienne Becker.

The British jewellery historian points to some great examples.

BLUE SAPPHIRE
4 carat

TSAVORITE
4 carat

ORANGE SAPPHIRE
6 carat

"Paloma Picasso works in a colour-block style – graphic and strong. Her style is a reflection of powerful femininity, quintessentially 1980s."

"Paloma Picasso works in a colour-block style – graphic and strong. Her style is a reflection of powerful femininity, quintessentially 1980s. She took full advantage of the wave of new coloured stones, using huge, intense pools of colour with stones like rubellite, drawing on her artistic heritage. Then there is Joel Arthur Rosenthal, known as JAR. JAR led the way forward by introducing previously disregarded coloured stones like spinels, garnets, coloured sapphires and more, in a multiplicity of different shades, to create the painterly, dégradé effect that is one of his most distinctive and influential signatures. JAR uses coloured gems as an artist uses paint. He has created a whole new emotive language of colour, introducing a new era of artistry, driven by cultural and historical references. Colour is and has always been an essential part of Bulgari's inimitable Italian style; the company has demonstrated a headstrong way with colour – exuberant and audacious in its mixing of contrasting colours with unique flair – as a counterpoint to the Roman architectural rigour of its designs. For Chopard, colour equals glamour – for Wallace Chan, the coloured stones are miracles of the natural world, waiting to be discovered, waiting to bring joy, life and meaning to the art of the jewel. Mikimoto have slowly and steadily increased their use of coloured stones, but in a very Japanese, very subtle, measured, considered way. I love the way they coordinate coloured stones with colours of pearls, the way they select only the best stones to complement the quality of the pearls and the way they set stones alongside pearls to bring out light and lustre in both and thus tell the stories of their collections. Finally, it can be seen how Victoire de Castellane, for Dior Fine Jewellery, has moved from using huge chunks of coloured gems as the focal point of figurative, narrative jewels, to far more graphic, sophisticated compositions of colour and form."

Without a doubt, the rules for gemstone-set jewellery have been broken during the past century. Vivienne Becker says: "This brought more fantasy, more story-telling into gem-set jewellery, more textural interest and more expression. The artistry of gem-set jewellery was also fuelled by new technological processes and even new materials such as titanium, which was used to enhance and intensify the colours of stones. But it was the use of colour that enabled jewellers to express themselves with more freedom, bringing more variety and more individuality to fine jewellery in general. Design is far less dictatorial today thanks to the discoveries of new gems."

"Before art deco, before the huge influence of the Ballets Russes, blue and green – sapphires and emeralds – were never used together. Then, suddenly, it became acceptable to combine colours in arresting, modern compositions and striking contrasts. First red and black, green and black, and the monochrome years of the mid-1920s, then the all-white vogue of the early 1930s, influenced by Hollywood's Silver Screen."

Vivienne Becker also believes that changing ideals of femininity through the past 175 years have been charted through the changing use of gemstones. "Women have become much more adventurous when it comes to designing, choosing and wearing jewellery. They are less driven by considerations of intrinsic value. They love colour, the way it fits in with clothes, personal style and what it says about the wearer. Coloured stones have enabled more story-telling in jewellery."

Another important factor is the fusion of fashion and jewellery which has taken place since the 1990s, when some of the largest fashion brands entered the fine jewellery arena.

LAGOON TOURMALINE
13 carat

LAGOON TOURMALINE
3 carat

"The influence of fashion on the use of coloured stones, the synchronicity of that fashion influence and the discovery of new stones have changed the look of fine jewellery dramatically. Perhaps the biggest, most radical change has been in high jewellery. Before the 1970s or 1980s, high jewellery was generally limited to very traditional designs, clusters with a central ruby, sapphire or emerald, surrounded by diamonds. High jewellery had to be classic, in order to ensure that it would be passed down through generations. But then with the beginning of the new millennium, with huge new wealth around the world and new coloured gemstones, high jewellery suddenly burst into creativity and originality, into complex narrative compositions, painterly colour schemes, fantasy and freedom," says Vivienne Becker.

The jewellery historian ends her account of the influence of newly discovered gemstones with one of the most recent 'superstars' to emerge on the market: the paraíba tourmaline. "I well remember the first buzz about paraíbas. That it was a colour never seen before in a gemstone, alive, vibrant, neon-blue, electric, like turquoise filled with light. It was and still is thrilling, a true natural wonder. Paraíba tourmalines opened jewellers up to even more expressive uses of colour, and this in turn opened up a whole new market, a new clientele eager for more exciting, less traditional jewellery. I also believe that the paraíba kick-started or at least fuelled the growing interest in the origins of gemstones, as this unprecedented colour of tourmaline came from one particular mine, in one particular area of Brazil. This added to the allure, tapped into the preoccupation with rarity and exclusivity. Paraíba tourmalines awakened us to the miracle of coloured stones. Then the discovery of paraíba-type tourmalines in Africa kept the market alive, stimulated the taste for its vibrant colour and also strengthened our interest in how and where these gems and minerals were formed – the geological relationship between the continents of South America and Africa, billions of years ago."

LAGOON TOURMALINE
6 carat

LAGOON TOURMALINE
11 carat

LUXURY BY NATURE

The explanation for Constantin Wild's love of gemstones is quite straightforward. "If a product is man-made, it is invariably an expression of the fashions at the time of its making. Contrary to that, gemstones are the results of Nature's moods millions of years ago. Each gem is unique – none are repeated. So, if I let an exceptional gemstone go, I can be sure I will never find another like it, regardless of the money that changes hands. Of course, to a certain extent I understand the attraction of luxury watches and cars, but as man-made mass-produced products they will never give me the unique emotions that gemstones do," he points out.

"I LIKE DIAMONDS, BUT LOVE GEMS. GEMS HAVE MORE SATURATED COLOURS, MORE INDIVIDUALITY AND MORE PERSONALITY."

Constantin Wild

COLOUR IS THE MAIN ATTRAC TION

COLOUR PERCEPTION AND GEMSTONES

By Çiğdem Lüle, PhD, FGA, GIA GG, DGA

The concept of colour is an extensively researched subject. Humans' colour perception is shaped by cultural practices. Previously, anthropologists theorised that all cultures see and name colours rather similarly. However, there is evidence that, although a degree of uniformity exists, people actually describe colour differently based on their social experiences. In other words, people perceive colour within a generalised context. For example, the colour pink was often correlated to 'rose'. In fact, while the word pink first appeared in 17th-century texts, most Eastern languages, regardless of their origin, used the word rose, not pink, as the colour description. The word pink (or its equivalent) was adopted much later, in the 19th century. Orange is another great example. Ancient Egyptians used the mineral realgar as the orange pigment. Undoubtedly, the colour orange has existed for millennia in art work and textiles, yet the use of the word orange in the English language first appears at the beginning of the 16th century. Prior to that, it was described as saffron, but even earlier descriptions of orange-coloured objects such as a robin's breast or orange hair were previously described as red. While a Mediterranean fruit took hundreds of years to be used as a colour analogue prior to the 20th century, a simple social media post may now bring anything unconventional to the world's attention, providing a brand new 'experience' instantly. Today, the constant expansion of technology and global communication has made information readily available to everyone interested.

Colour perception is puzzling not only to social scientists but also to neuroscientists. In essence, while we learn and use colour descriptions in our daily lives through experience, how colour is processed by the human eye and brain is not yet well understood. It is simply not known if two individuals positioned next to each other and viewing the same object will perceive its colour the same way. This particular notion has motivated scientists to further research. Consequently, colour-related industries such as print, paint and textile control this issue through the development of structured, comparison-based colour systems. The use of practical comparison-based colour standards or catalogues has been found to reduce confusion among observers.

Human nature tends to appreciate beauty and rarity regardless of the era or culture. Apart from diamonds and pearls, virtually every other gemstone that is deemed valuable is a coloured gem. Modern coloured gemstone grading and valuation practice demonstrates that the most important feature of a coloured gemstone is its colour.

Humankind's attraction to gems and minerals precedes recorded history and may have many different reasons, some of which we can only speculate about today. The allure of gemstones stems from the fact that they are the rarest and purest form of minerals. Aesthetic notions aside, minerals have been a source of numerous raw materials, including pigments that are used in fundamental industries such as textile and painting. The oldest examples are ochre, derived from haematite to create shades of yellow, red and brown, and metallic grey, extracted from graphite. Mineral-based pigments have been so valuable that the blue pigment ultramarine, obtained from lapis lazuli, was considered as precious as gold until the 19th century.

COLOUR IN OUR DAILY LIVES

By Katinka Champion, AOCA, CMG contributor

Most of us have a favourite colour, whether we realise it or care to admit that we have one. If you ask young children about their favourite colour, they will usually blurt out an answer or sometimes even reply with multiple colour favourites.

Why do we prefer one colour to another? Did the love affair start at an early age when you received your first set of colourful crayons? Or was it the 'British racing green' toy sports car you received on your fourth birthday? Or the amethyst purple bauble ring you were given in your loot-bag at a friend's birthday party when you were just six? The list could go on, but only you can truly know what positively valenced event, or object, triggered your colour preferences.

These positive emotional experiences with a specific colour are likely to evolve into our colour preferences, which get embedded into our psyche – we carry their memories with us. Sometimes we don't even know where, or when, the connections were made. They just surface, from time to time, to help us make colour decisions in our daily lives.

If we look up the meaning of each spectral colour on the colour wheel on the Internet, we are bombarded with a list of emotions and facts. For example, if we were to google the colour red, words that might be listed are passion, desire, love, power, danger, anger, rage, violence – to mention just a few. The colour can also increase our appetite, as it revs up our metabolism, increases our respiration rate and raises our blood pressure.

The meaning of a particular colour is never a simple, one-answer response. Many cultures assign different symbolic meanings to individual colours. In Chinese culture, for example, white represents death and is also the colour of mourning for a funeral, whereas in Western culture the colour white denotes purity of spirit, and peace, like a traditional white wedding dress, and a pure white diamond engagement ring – two diametrically opposed emotions for the same colour, but in different cultures.

I'm truly no longer interested in the textbook answers to the emotional meanings of colours which are 'only a Google away', like green with envy and red with rage.

"Fall in love with the colours that speak to you."

I am more interested in the collective psyche, like when five women in one office show up on the same day at work wearing exactly the same hue of red suit jacket that wasn't even popular that season. What truly resonates with me now are the meaningful stories that colour evokes in us all, that are unique to our own life histories and cultures.

Colour System Company Pantone first launched its 'Color of the Year' (COY) concept in 2000. The process they go through is intentionally shrouded in secrecy. What Pantone has made public is that twice a year, in a European capital, their colour specialists meet secretly with an anonymous group of trend forecasters and international colour experts from a vast range of design industries. After two days of presentations on a themed topic, and lots of brainstorming and debate, they choose a colour for the following year. Pantone says the annual colour is selected based on "what is taking place in our global culture at any given moment in time".

'Pantone Classic Blue' was chosen to herald the start of a new decade. Pantone describes this "deep shade as a reassuring presence instilling calm, confidence and connection. A timeless and enduring hue elegant in its simplicity", which I'm sure will permeate well into the decade, as the hue is viewed as "solid and dependable, suggestive of the sky at dusk".

In Constantin Wild's world of gems, 'Classic Blue' translates into several timeless gemstones, like the confident, classic sapphire, the vibrant violet-blue tanzanite and the mysterious Sri Lankan purple-blue spinel. At Constantin's atelier the 'Classic Blue' dusk sky has no limits – the possibilities are endless!

Colour trends are often cyclical with slight nuances that make them fresh and current. Sustainability of our resources is a key factor in our new decisions, and this is where the concept of 'heirloom' pieces in jewellery plays a key role – in much the same way it has in the past with some traditional cultures. Think more in terms of acquiring jewellery with magnificent, coloured, cut gems to pass on to future generations.

Fall in love with the colours that speak to you, and curate well designed, bespoke jewellery pieces that can move with the times. Do your part in making fashion more sustainable by creating heirlooms the Constantin Wild way with masterfully cut natural gemstones.

BLUE GEM CLASSICS

SAPPHIRE, 8 carat

VIOLET-BLUE TANZANITE, 16 carat

PURPLE-BLUE SPINEL, 12 carat

OPINION...

A conversation between Constantin Wild and Katerina Perez

by Katerina Perez

Gemstones are more than their chemical components. They are a snapshot of the history of the earth, a conduit for colour and style, and the backbone of an industry that meticulously considers colour, cut and clarity, among countless other factors that separate the beautiful from the sublime. Ask an experienced gemstone merchant to gaze upon a stone and you will discover that he or she can read its facets like chapters in a book. Many demonstrate a remarkable sixth sense, quickly outlining the nature of a stone and discussing its characteristics. Nothing surpasses the moment when a truly spectacular stone lands on the desk of a gem expert: witnessing the immediate fizzing excitement, the stone being carefully turned and viewed from all angles, the hint of a smile. It's a testament to the passion that goes hand in hand with a lifetime in the gem trade.

My friend, Constantin Wild, embodies this natural affinity with gemstones. He is adept at selecting the top one per cent of stones, destined for high jewellery collections and private collectors, as well as the remaining top three per cent of all finds that are used to adorn fine jewellery creations. To hear him talk about gemstones is to understand how serendipitous their existence truly is. Saturation of colour, so vital for beauty and character in aquamarines, imperial topazes and tourmalines, for example, is so delicate a scale that becoming too saturated can diminish a gem's brilliance and attraction.

Rare occurrence in Nature is another factor that boosts the value of a gem. Next, Constantin Wild may discuss the relationship between value and size, and how the cost of a gem is calculated by its price per carat – the bigger the gem is, the higher its price. As some minerals only occur as small crystals; a large gemstone cut from one of these natural origins can have a higher price per carat, for the same quality, than that of gems cut from larger crystal structures. Examples include rubies, garnets (green demantoid, green tsavorite, orange mandarin garnet) and tourmalines like paraíba and blue indigolite.

And while inclusion-free, flawless gemstones are often seen as the pinnacle of Nature's achievement, Constantin Wild reminds me that Nature doesn't do flawless – Mother Earth is simply asking us to look more closely. Even a 'flawless diamond' isn't flawless when placed under more than 10x magnification. Inclusions are birthmarks which tell us a lot about the genesis of a gemstone. What they shouldn't do, however, is interfere with the overall appearance and limit the brilliance of a stone. Eye-clean, meaning inclusions that cannot be seen with the naked eye, is suitable for jewellery purposes yet, in fact, some coloured gems increase in rarity and value with certain inclusions, such as demantoid garnets and their horsetail signatures.

This chapter is a retelling of my gemstone conversations with Constantin Wild and my attempt to affix his opinions, passions and experiences to the page. I hope you will see, as I do, that his knowledge could create many forks in the road, leading us off in different directions to the mines of Sri Lanka, the plains of Africa and the ancient riverbeds of Madagascar and Thailand.

AQUAMARINE, MOZAMBIQUE
44 carat

GREEN BERYL, BRAZIL
72 carat

BERYL

A colourful family

EMERALD, COLOMBIA
6 carat

RED BERYL, UTAH
0,7 carat

This family of gemstones is full of strong characters; the verdant green of emerald, the feminine blush of morganite and the sweeping blues of aquamarine. "The beryl is a hard stone that allows for a good polish and lustre – a fantastic stone to collect," Constantin Wild says, with the matter-of-factness of someone who has handled some of the world's finest specimens. He continues: "Emeralds are perhaps the most sought after, with that classic shade of green, but there's also aquamarine, morganite, heliodor and, finally, the very rare red variety." Whole books have been written on the ancient and coveted emerald, with its inner world of inclusions that spread like ivy tendrils through the stone. And while it may have the loudest voice of the beryl family, Constantin Wild is equally drawn to its no less beautiful sisters, such as aquamarines. "The more saturated the blue, the more valuable it becomes. However, it is the lighter blue shades that are more frequently discovered. Saturated aquamarines have always been the most desired and expensive ones. For me, the origin is of little importance; it is only the pure beauty and colour saturation of the aquamarines that matter to me," Constantin Wild explains.

When discussing aquamarine, it's likely you will hear the name of the Santa Maria mine in Itabira, Minas Gerais, Brazil, which has been closed since the mid-20th century. Constantin Wild has handled only two such aquamarines, which demonstrates just how rare these gems are. He says: "Around the turn of the 21st century, Santa Maria aquamarines suddenly re-emerged in the market, but they were dubbed 'Santa Maria Africana' to indicate their origin in Mozambique. They have a colour and appearance like that of the Santa Maria aquamarines from Brazil. Unfortunately, in today's market, the attribute 'Santa Maria' is casually applied to any kind of aquamarine unlinked to source or quality, which has caused the once important designation to lose some of its sparkle."

The feminine morganite has been recognised for its hues of pink and peach but has only been known for less than a hundred years. Constantin Wild recounts the story: "It was in 1911 that, on the suggestion of the New York gemmologist G. F. Kunz, the pink variety of beryl was ennobled to the status of a gemstone. To honour the banker and mineral collector John Pierpont Morgan, it was given his surname, under which it is known today: morganite."

As with so many conversations about the beryl family, my conversation with Constantin Wild also touches on the phenomenally rare red variety of beryl, sometimes controversially called 'red emerald'. "The dark red colour is attributed to manganese ions and it is extremely rare. It has only been reported in a handful of locations, including the Wah Wah Mountains in Beaver County, Utah. A red beryl with a cut weight of more than half a carat could be considered a once-in-a-lifetime gem."

MORGANITE
112 carat

AQUAMARINE
70 carat

PARAIBA TOURMALINE,
NIGERIA, 4 carat

PARAIBA TOURMALINE,
MOZAMBIQUE, 2 carat

PARAIBA TOURMALINE,
BRAZIL, 3 carat

TOURMALINE

Nature's rainbow

GREEN TOURMALINE
24 carat

LAGOON TOURMALINE
38 carat

RUBELLITE
46 carat

CHROME TOURMALINE
15 carat

CANARY TOURMALINE
14 carat

LAGOON TOURMALINE
6 carat

"Tourmaline has one of the widest colour ranges of any gem species, occurring in various shades of virtually every hue." That's how Constantin Wild begins when I ask for his thoughts on tourmaline – Nature's rainbow material – which can be found worldwide, but especially in Brazil, Nigeria, Namibia, Mozambique, Pakistan and Afghanistan. The Spanish conquistadors of the 16th century are said to have stumbled upon tourmaline in Brazil, mistaking its colour for the verdant green of emerald. This confusion was a constant companion of tourmaline until it was separated as a distinct mineral species in the 1800s.

Constantin Wild says: "With common pink, red and green tourmalines, the carat price remains largely the same, regardless of whether a stone weights five carats or twenty. As rough crystals usually occur in various sizes, the carat price depends solely on colour: pink or green."

Tourmalines are usually flawless or eye-clean, except for red tourmalines (rubellites), where inclusions are generally tolerated due to the simple fact there are no flawless red tourmaline crystals in Nature. One could spend a lifetime observing tourmalines, from harmonious shades of green, feminine pink, fiery red rubellite, masculine blue indigolite and greenish-blue lagoon tourmaline – a fashionable hue for the 21st century.

Of course, there are some tourmalines of note, specifically paraíba tourmaline and canary tourmaline, which exhibit neon blue and zingy lemon colours, respectively. "The extremely rare paraíba tourmaline was first found in 1989 near the village of São José da Batalha in Paraíba State, north-eastern Brazil, and later in Nigeria and Mozambique," Constantin Wild explains. "Finding the same gem varieties in Africa and South America is common, as both continents were once linked together as one super continent called Gondwana, which separated about 180 million years ago. Paraíba tourmaline is typically a small neon blue or greenish-blue gem. Most tourmalines get their colour from traces of iron, manganese, chromium and vanadium, but the paraíba has traces of copper, and often manganese, that interact to enhance its beauty."

For African and Brazilian tourmalines, price is determined by the saturation of colour, the beauty of the neon-blue nuance, and the absence or near absence of inclusions. Brazilian Paraíbas generally fetch higher prices because their colour saturation is usually more intense than their African sisters. "Production of Brazilian paraíba tourmalines, which was rather small in size and quantity, stopped around the turn of the century, thus ensuring that prices for quality stones in the one-carat-upwards range have reached a stratospheric price per carat," Constantin Wild explains. "The mines are exhausted and, today, paraíba tourmalines from Brazil are almost exclusively available on the secondary market, where they fetch high prices."

As for African paraibas from Nigeria and Mozambique, Nigeria has not produced much in the past decade and Mozambique production is also very limited. The result? Paraíba tourmaline has 'surreal' prices that reflect the insatiable demand for its fresh, swimming-pool blue – a colour that, according to Constantin Wild, "deserves the hype it is given".

Traverse the tourmaline rainbow and you will find canary tourmaline, a zesty lemon variety mined in the Lundazi District of eastern Zambia, close to the border with Malawi. Constantin Wild shares the story: "Mining started in 1983, but only since the turn of the 21st century has this gem-quality tourmaline reached the market and secured the name 'canary'. This small mine produced only a handful of gems. Whereas paraíba tourmalines get their colour from copper, the canary colour comes from traces of manganese. Today the mine is exhausted and closed."

Canary tourmaline has a very lively colour, similar in character to that of mandarin garnets or paraíba tourmalines. As it is the only true lemon-yellow gemstone, demand for it in high jewellery palettes is increasing. "The spring-like vivid yellow of these tourmalines is clear and pure, perhaps with a very fine hint of green," Constantin Wild explains. "I like both varieties – the ones with a greenish hint and those without; they are equally valuable. Canary tourmalines of more than five carats are extremely rare."

Tourmaline is the chamaeleon of the gem world: colourful, adaptive and versatile for jewellers. It is impossible not to fall for at least one of its myriad colours, especially those with the elec-tric energy of this mineral species' rarest members.

Pink tourmaline – An empress's pillow

Sometimes an individual's love for a particular gemstone can morph into a fascinating and historic obsession. Take for example the empress Dowager Tzu-Hsi (1835-1908), the last empress of China, who bought almost the entire production of gem-quality pink tourmaline from mines in California, USA, during the final years of her life. The imperial Chinese government is said to have imported 120 tons of tourmaline rough, which was later used for jewellery and carvings. Perhaps the most unusual carving was a pink tourmaline pillow, upon which the empress was laid to rest.

RED TOPAZ
4 carat

IMPERIAL TOPAZ

Wild emperor

Constantin Wild is one of a handful of gemstone connoisseurs who have witnessed the most regal and sophisticated heights of topaz. Topaz is most commonly associated with shades of blue, which are very rarely natural, and it is a gemstone used in commercial jewellery collections the world over. By contrast, the golden, peach, orange and red tones of imperial topaz are the remit of collectors, distinguished not only by their rarity but also by their sunset-like beauty, bringing a natural warmth to high jewellery creations.

Imperial topaz was first discovered in the late 18th century by the Portuguese and it swiftly became fashionable among the royalty and aristocracy of Europe. However, when Constantin Wild first purchased several specimens from Brazil in the 1980s, this stunning mineral was only known to a small circle of people. It had lost its starry status to the passage of time.

Constantin Wild says: "For a long time, the noble topaz was only for connoisseurs. In recent years, this valued gem has enjoyed a renaissance, the most beautiful variety of which, the Brazilian imperial topaz, has become extremely rare."

In Constantin Wild's collection are some of the rarest, most saturated topaz colours: deep orange, pastel pink, hot pink and reddish orange, each with a characteristic golden light that appears to emanate from deep within the stone. A combination of pink and gold hues is considered the most valuable. The price of imperial topaz from Brazil can vary several hundred-fold depending on the depth and vivaciousness of the colour. And, unlike colourless topaz, heat or irradiation treatments do not have the same effect on imperial topaz, which means the scarce material on the market is less likely to have been altered by man-made science.

Why 'imperial'?

Although the primary source for imperial topaz is considered to be Minas Gerais in Brazil, the 'imperial' moniker was given in honour of the Russian monarchy following the discovery of topaz in the Ural Mountains in the 19th century. This imperial topaz had a golden-reddish hue, sometimes likened to that of sherry, although true red imperial topaz is considered extremely rare indeed.

PINK TOPAZ
33 carat

IMPERIAL TOPAZ
8 carat

IMPERIAL TOPAZ
11 carat

IMPERIAL TOPAZ
4 carat

IMPERIAL TOPAZ
18 carat

IMPERIAL TOPAZ, BICOLOUR
15 carat

DEMANTOID GARNET

1 carat

PYROPE GARNET

14 carat

GARNET

Queen of reinvention

Our conversation veers from the rarest of red beryls to the more widespread red garnet. "For many years, the garnet was held to be a somewhat common red stone and certainly not regarded as anything special. It was simply brownish-red pyrope garnets that were set in 'grandmother's jewellery," Constantin Wild explains. Yet the humble garnet enjoyed a kind of renaissance – a period of reinvention – in the second half of the 20th century, when "the jewellery world came to realise that the garnet group of gemstones is in fact very colourful", says Constantin Wild. During this time, jewellers were excitedly exploring the rose pink and scarlet shades of rhodolite garnet, the vibrant greens of tsavorite and demantoid garnet, and the fiery orange of mandarin garnet, which Constantin Wild describes as "sparkling, lively and beautiful".

He continues: "As garnet crystals tend to be small or very small, all the tsavorites, mandarins and demantoids become exponentially more valuable with size. A curious fact is that the demantoid is more valuable with inclusions – known as horsetails – than without, because inclusions are considered beautiful birthmarks. Clients are well advised to look especially for this particular inclusion when buying demantoids. Today demantoids over one carat in weight are considered spectacularly rare, especially those with a saturated chrome green colour and fine horsetail inclusions."

From overlooked and underappreciated to bright sparks in a gem collector's palette, the garnet is the queen of reinvention and revival.

GROSSULARITE GARNET
13 carat

PYROPE GARNET
14 carat

MANDARIN GARNET
17 carat

TSAVORITE GARNET
7 carat

FIRE OPAL, MEXICO
32 carat

OPAL

Fire within

In the early 1880s, my great-grandfather was one of the first importers of Australian opals to Europe," Constantin Wild begins as we browse fire opals, white opals and black opals with the most magical flashes of rainbow shades. "Personally, I like opals when they are colourful with a full colour play of red, orange and blue," says Constantin Wild. "I also like boulder and black opals. The latter are rare and much sought after. For a long time Australia was the main producer, but now opals with orange to reddish undertones, with or without colour play, are also being found in Mexico. Some of these are transparent enough to be faceted as regular gemstones. White opals have also been discovered recently in Ethiopia."

So many of my conversations with Constantin Wild become travelogues, referencing different corners of the globe and countries that, albeit diverse, are united by the gems that lie beneath their soils. In this instance we project ourselves to Mexico, specifically the Laureles mine in Yalisco, Sierra Madre, to discuss what is presumed to be the largest fire opal in the world, discovered in 2006. Constantin Wild says: "Back in the time of the Mayas and Aztecs, the precious 'Quetzalitzlipyollitli', or 'stone of the bird of paradise', as they called it, was coveted. They believed that the fire opal, with its vivacious sparkle, came from the waters of paradise. For that reason, this stone is named after the Aztec goddess of water, 'Chalchiuhtlicue' (pronounced Chal-chi-utl-ikwe)."

FIRE OPAL, MEXICO
23 carat

FIRE OPAL, MEXICO
24 carat

FIRE OPAL, MEXICO
32 carat

FIRE OPAL, MEXICO
15 carat

FIRE OPAL, MEXICO
'CHALCHIUHTLICUE'
132 carat

TANZANITE
21 carat

TANZANITE PAIR
28 carat

TANZANITE

The purplish-blue gem of Kilimanjaro

The human fascination for gemstones can be traced back thousands of years, but that doesn't mean that we can't still be surprised by contemporary discoveries. A perfect example of that is the unearthing of purplish-blue zoisite in the foothills of Mount Kilimanjaro in 1967. It was swiftly named 'tanzanite' by Tiffany & Co. in honour of its geographic origin – and the rest, as they say, is history. "I like the colour because it is not simply plain blue. Its pleochroism is blue and purplish, and I like the particular mix it creates," Constantin Wild explains. "Some people think that the bluer the tanzanite is, the more sapphire-like it is – so a significant purple tone gives tanzanite an interesting change of character. I always go for the saturated colours when I choose which gems to sell – but I am always careful not to choose the over-saturated stones as these are less attractive." Unlike other stones, tanzanite has a price per carat that remains largely static, regardless of whether it is a five-carat or a 20-carat stone, which means it is the vibrancy and saturation of its colour that determines the value. There's an inherent romance to a stone that is found in just one geographic origin, almost like buried treasure. Fortunately, humankind was lucky enough to stumble upon the 'X' on the map, and today, gem traders like Constantin Wild have another shade of bluish-purple to add to their palette.

BLUE ZIRCON PAIR, CAMBODIA
82 carat

ZIRCON

Rhapsody in blue

There are parallels between the story of spinels and another underappreciated gemstone, zircon. "For a long time, natural colourless zircons were used as diamond imitations, so they acquired a bad reputation," Constantin Wild explains. The zircon was linked to diamond imitations through no fault of its own – the situation being compounded by the name given to synthetically man-made diamond-like imitations: cubic zirconia. In reality, natural zircon is a "remarkably beautiful and colourful" gemstone with a varied palette of yellow, green, red, reddish-brown and blue. Constantin Wild says: "Zircons have strong double refraction and are extremely lively and radiant. I own probably the largest pair of oval-cut blue zircons with a weight of more than 80 carats."

BLUE ZIRCON, CAMBODIA
80 carat

BLUE SAPPHIRE

8 carat

RUBY

8 carat

ORANGE SAPPHIRE

12 carat

PURPLE SAPPHIRE

5 carat

CORUNDUM

Strong and mighty

Beneath the umbrella of corundum lie two of the gem world's biggest superstars: ruby and sapphire. Both are surrounded in lore in their own way, from the royal flush of 'pigeon's blood' Burmese rubies to the deep satin blue of Kashmir sapphires. The finest examples of both these gems command extraordinary prices at auction. Constantin Wild provides the background: "Ruby is the red variety of the mineral corundum. Slight traces of elements such as chrome, iron, titanium or vanadium are responsible for the colour. At 9 on the Mohs scale, ruby has excellent hardness and is second only to diamond. Mozambique has recently been producing commercial and fine-quality ruby, but the most important thing to consider is always its colour."

Among the gemstones dealt in by Constantin Wild are rubies, blue sapphires and fancy-coloured sapphires. Red rubies and blue sapphires are, of course, amongst the best known and most desired gems, but the corundum family has the whole spectrum of colours available, from yellow and orange to pink, violet, purple, green and black. "I am an admirer of the violet, purplish and orange ones," Constantin Wild says. "I do like the traditional blue sapphires, but they are so much in demand that the prices make them nearly impossible to get hold of – so I would rather look for other rarities and beauties that are still affordable." That is what's so special about corundum; its public persona is dominated by two primary colours, yet just behind them there is a veritable rainbow of gems, each one ready to add soul to jewellery creations. Never before have supporting characters been quite so charming.

BLUE SAPPHIRE

4 carat

PURPLE SAPPHIRE

9 carat

PURPLE SAPPHIRE

6 carat

VIOLET SAPPHIRE

6 carat

RUBY

2 carat

PINK SAPPHIRE

7 carat

ORANGE SAPPHIRE

17 carat

YELLOW SAPPHIRE

26 carat

PERIDOT, PAKISTAN
105 carat

PERIDOT, BURMA
86 carat

PERIDOT

Moss and silk

In the mid-1900s, peridot deposits were located at 4,000 metres (13,120 ft) in the Kashmir region of India, which might suggest they should be sky-blue in colour! Of course, peridot is a moss-green gem and, from this region in particular, the gems are "of an incomparably beautiful colour and transparency", according to Constantin Wild. These gems are offered as 'Kashmir peridots' – utilising the moniker more commonly associated with sapphires to emphasise their beauty and rarity. Mining at these dizzying altitudes takes skill and patience. Tough weather conditions mean that only the summer months are suitable for it.

Within his collection, Constantin Wild has a 105-carat peridot from this region, which he describes as a "sure collector's piece". What else should we know about peridot? "Mogok in Burma is known to produce the best-quality peridot from two mines near the Barnardmyo villages: Pyaung Gaung and Zalat Thaung. Myanmar peridot is rare and of a warm, saturated, silky-green colour. It is often more attractive than peridot from other sources and thus quite valuable, approaching the per-carat values of top gemstones." As a striking example, consider this 86-carat cushion-cut peridot, which is a unique collector's piece.

AMETHYST
8 carat

AMETHYST
38 carat

AMETHYST

Purple of the Roman Caesars

Although amethyst is the most highly valued stone in the quartz group, it brings far more than just its value to the international gem market. Firstly, Constantin Wild and I discuss its colour – a true, regal purple – which is caused by the presence of iron. "The most important deposits are in Brazil," Constantin Wild explains, "specifically 'Palmeira' amethysts from the state of Rio Grande do Sul in the south and 'Marabá' amethysts from the northern state of Pará." Amethyst also has a fascinating history, with ancient Greek legends linking the gem to Bacchus, the god of wine, because of its grape-like colour. Amethyst also has centuries of connections to the Christian church and is one of 12 gemstones associated with the 12 apostles, the disciples of Jesus Christ. This lofty religious significance is likely one of the reasons why amethyst appears in the crown jewels and coronation regalia of British royalty. "Amethyst is one of my favourite gems. It's woven into so many mythical stories and has been associated with drunkenness, love and intelligence throughout history. Amethyst may come from a modest gem family, but its vibrant vitality is not to be underestimated."

BLUE SPINEL, SRI LANKA
13 carat

RED SPINEL, TANZANIA
5 carat

SPINEL

Undervalued beauty

"It wasn't until about 150 years ago that spinel was recognised as a mineral in its own right," Constantin Wild explains as our conversation flows to spinel. "Lo and behold, some well known 'rubies' in royal collections were later proved to be spinels, among them the Black Prince's Ruby in the Imperial State Crown of the British crown jewels and the pear-shaped stones in the crown of the Wittelsbachs of Bavaria."

Traditionally, spinels were found in the gemstone-bearing mines of Sri Lanka but, more recently, too in excavations at Tunduru in Tanzania and Ilakaka in Madagascar. They occur in various shades of pink, red, blue, green, yellow, grey and violet. Constantin Wild continues: "The spinel is beautiful because of its high brilliance, which is due to its hardness and high refractive index." So why do those with decades of experience still consider the spinel to be one of Nature's most undervalued gems?

To understand this, one must traverse its history and consider the unfortunate meddling of man. Constantin Wild explains: "Unfortunately, for a hundred years or so, spinel was not regarded as 'genuine'. At the beginning of the 20th century, the spinel became the first gemstone that could be produced synthetically, with the Verneuil process. The market was flooded with synthetic spinels."

Fortunately, the market for this fantastic gem has recovered with education, and today's jewellery lovers have embraced its natural colours, from soft pink to strong red, blue to grey and grey to purplish. Constantin Wild is on his own personal mission with spinel too. He says: "Currently, I am opening my heart to the purplish-grey earthlike tones of spinel. I don't wear a lot of jewellery myself, but this appeals to me so much that I need to own a ring with just such a shade!"

PINK SPINEL

5 carat

VIOLET SPINEL

8 carat

RED SPINEL

8 carat

PINK-RED SPINEL

5 carat

SUP
ER
LATI
VE
JEWE
LLERY

SUPERLATIVE JEWELLERY

by Nina Hald

Original. Lavish. Romantic. Sensual. Superlative jewellery – *haute joaillerie* or high jewellery – is all about gems set in inspired designs. Gifted individuals create unique designs that add further radiance to gems. At its best, high jewellery is nothing short of spectacular. Often, designers will say that they build a design, for example that of a necklace, around a chosen gem. Some even proclaim that 'the stones are the stars' on select, exclusive creations. And they may well be. When a designer is struck with a divine idea for a piece of high jewellery, its realisation must still await the sourcing of the perfect gems.

Historically, jewels were a symbol of wealth and power rather than of style. Predominant style idioms and society perceptions were reflected in the arts, such as architecture, music, painting and clothing style, and these naturally also found reflections in jewellery. Gems were used for ceremonial objects such as dress swords, for seals and signets to mark property and letters and for religious reasons, lithotherapeutic (medicinal) functions and dynastic purposes. Entitlement to the throne of a country was affirmed by regalia such as gem-encrusted crowns, sceptres and rings. Set in jewellery and other costly objects, gems projected status – and for those reasons, they are still displayed in collections at museums, in curiosity cabinets and treasure chambers.

During the last 150 years, jewellery has become increasingly linked with feminine fashions. In fine and high jewellery, gems set in virtuoso designs, imagined and created by talented designers, represent the very best of hand-made craftsmanship. Wearing such wonders on the body lends another dimension to the beauty of the gems – and to us as humans. The sumptuousness of gems is as close to eternity as any of us can hope to get. People in general tend to forget how long it takes to find and cut gems, but as Constantin Wild phrases it:

"We all like to own something unique which makes us stand out and something that fits our style. This is best accomplished by wearing coloured gems. Gems are a solid part of the Earth, considered to be the most valuable and beautiful manifestations of the inorganic – we as humans like to own part of this solid world, which is as unique as we are ourselves."

High jewellery creations are designed around gems, as time-honoured traditions dictate; the creations are fragments of pure ecstasy, fuelled by admiration and desire. From the most traditional to the most extravagant, Constantin Wild gems are ambassadors of jewellery virtuosity and confident elegance. Gems light up the night with brilliance and colour, extravagance and sophistication, in a timeless sparkle that will last for ever.

Photo: Fred Leighton

TSAVORITE

5 carat

TANZANITE

36 carat

NEW SUPERSTARS

Decades ago, Fritz Wild took his young son Constantin aside to show him a radiant green crystal, still in the rough. The colour was absolutely mesmerising. The boy, by then well versed in the family business, -instantly knew that he had never seen the likes of it in the family offices, so the very first question that stumbled out of his mouth was: "What is it called?" "Alas," came the response, "it has not been named yet, but I am sure we are going to see many more of them in the future," said Fritz Wild.

It was a tsavorite – a green grossular. And Fritz Wild was indeed right; the company in Idar-Oberstein later handled many tsavorites, cutting them to perfection, after which they were sent on their merry way out into the world to be set in jewellery. Tsavorites were discovered in 1967 by the Scottish geologist Campbell R. Bridges (1937-2009), who worked as a consultant for Tiffany & Co. In 1974, the gemstone was named after Tsavo West National Park by Henry Platt (1924-2015), then president of the legendary American jewellery company and great-great-grandson of founder Charles Tiffany (1812-1902). At about that time, Tiffany & Co. also (re-)named the violet-blue zoisite variety discovered by the Portuguese tailor and part-time gold prospector Manuel De Souza: tanzanite.

From fathers to sons, the legacy of gems and the love of colours have been passed on countless times in the company of Constantin Wild. The tsavorite is but one example. From time to time, new gems have appeared on the global jewellery stage – some even stealing the spotlight and becoming 'staples' thanks to their glorious colours.

Hauyns. Tsavorites. Tanzanites. Canary tourmalines. Mandarin garnets. And of course, the most recent superstar of the gemstone business: the paraíba tourmaline. These relatively newly discovered gems have upended and transformed the jewellery industry with their superlative colours by giving designers many more nuances to play with – they continue to inspire jewellery designers and goldsmiths to make fabulous creations. If Marilyn Monroe had known them, those are the ones she would have been singing about today.

Some of the most gorgeous gems have been discovered only during the past 50 years, and Constantin Wild, having been born into one of the most important gemstone-trading families, has witnessed their launch on the market. His father saw the tanzanite and tsavorite brought into the market – Constantin Wild has observed the introductions of mandarin garnet, demantoid garnet, paraíba tourmaline and canary tourmaline, even playing a part in most of them. In addition to these all-new discoveries, Constantin Wild has also taken part in new finds in Tanzania: blue moonstone and pink and red spinels.

CANARY TOURMALINE

10 carat

A FANTASY GIFT

For almost a century, each Christmas season, the originally Dallas-based retailer Neiman Marcus has offered extremely luxurious gifts not intended for normal sale. Some of the highlights of this tradition have included a submarine trip to the deepest part of the ocean, James-Bond-styled cars, bespoke perfumes and personalised Maseratis. While these are sometimes announced months in advance, each December brings to light some eight Fantasy Gifts, as they are called. In 2014, one of these Fantasy Gifts was a trip to Idar-Oberstein with the Connecticut-based jewellery designer and photographer Monica Rich Kosann. Final destination? The Constantin Wild company, where a gemstone was to be selected. As part of this experience, the designer would suggest designs overnight for a locket featuring the particular gemstone chosen by the winner, with delivery to him or her at a later date. The Fantasy Gift was set at 100,000 dollars, and it sold within minutes of being announced - to a mother and her daughter, Marilyn Pape and Shelley Bennet. Together, they decided on a 15-carat purplish-red cuprian tourmaline in magenta and shocking pink from the Paraíba mine, a choice from Constantin Wild's collection, around which a golden locket was made with a photograph of the two inside, in memory of the once-in-a-lifetime trip. The cuprian tourmaline was framed by matching melee of Sri Lankan spinel in various shades of purple and shocking pink and hung on a golden link chain.

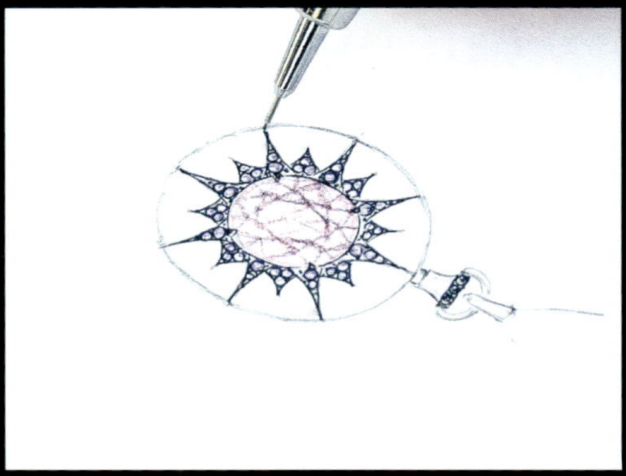

MANDARIN GARNET

17 carat

NAME THAT MANDARIN

One of the more recent stories was the naming and marketing of mandarin garnets in which Constantin Wild has taken part. The mandarin garnets were first discovered in northern Namibia about thirty years ago.

"I was immediately thrilled when I saw their orange colour, because it was clear to me that it would work very well with the fashion colours at the time. It was a new, fresh and juicy colour,"

Constantin Wild recalls. The mandarin garnets made their way on to the market in 1991, and as Constantin Wild had sensed, their clear vivacious colour fit perfectly with the times. "As I remember it, there was a dispute about what to name them. One person wanted to call them hollandine; another suggested naming them after the Kunene Valley, where they had been found – but when I came on board this unofficial naming committee, I and a few others suggested that we simply follow the mandarin-orange colour," and that is exactly what happened.

The first finds from Namibia were rather small in size, mainly melee; the first mines were quickly exhausted too. "One of the largest mandarin garnets, likely the second-largest ever found in Namibia, was about 15 carats after cutting, and I immediately – and unfortunately – sold it. Today, it would be almost priceless," says Constantin Wild with regret. And how much he regrets having parted with that treasure can clearly be heard in his voice. However, with the grand overview of high jewellery being made today, Constantin Wild is supremely happy to see so many coloured gems set in creations, compared with how it was just two or three decades ago. "Gems are finally and well deservedly taking centre stage. The designer and artist Paloma Picasso was one of the first to pick up on the new colours. Some of the large rings she designed for Tiffany & Co. feature single coloured gems in extraordinary presentations. When I look at Paloma Picasso's pieces, it is obvious to me that her thought process begins with the large rings. From them, she moves on to matching earrings and then complete, entire jewellery suites with coloured stones," Constantin Wild says.

Naturally, for Constantin Wild it is a particular pleasure to sell a larger suite of stones – for example sixteen gems – to one of the large jewellery companies. "I know they will make something fantastic out of each suite. Sometimes a jewellery designer will be on the lookout for a particular gemstone, but then his or her eye is caught by one or more gems in my collection, and he or she goes crazy for them instead," Constantin Wild explains. For him, it is both intellectually stimulating and a lot of fun to open the boxes of rough and cut gems in the company of designers from the major brands. "We compare colours while discussing matches and brilliancy, as the table becomes a creative mess of gems, styles and ideas. Good art directors assemble combinations that I would never have thought of. With their eyes for this, they can literally – on the spot – create something new, and from that they develop new concepts. To witness designers mixing and blending gems, shifting them around in order to achieve new colour combinations, is an amazing experience every time," says Constantin Wild.

PARAIBA TOURMALINE
8 carat

NATIONAL PREFERENCES

At Constantin Wild, drawers are kept well stocked with calibrated stones, as the company has concentrated on classic shapes like oval, cushion, octagonal and round. "In the 1980s and 1990s, there was more demand for the so-called fancy shapes, and of course we could supply them, but I prefer timeless cuts, as they are the most enduring in style. Since I became the sole owner of the company in 1996, we have not done many fancy cuts," Constantin Wild explains.

For the past twenty years, Constantin Wild has sold a great many centre stones paired with colourful melee for his Asian clients. "When I entered the Japanese market in the mid-1990s, I noticed their proclivity towards high-quality gems in pastel colours, such as aquamarine and imperial topaz, between 3 and 5 carats in size, which we supplied at the time in calibrated sizes and usually oval in cut. The Japanese market was also the first that 'took' to paraíba tourmalines because they loved the fresh neon-blue nuances. From my observation, the Italians and Americans usually prefer strong colours and larger stones – in the 10-15-carat size range; the European markets orient themselves towards bright colour combinations. When the Chinese entered the gemstone market, they loved large-sized gems, in everything from pastels to strong colours. As long as the gems were sizeable, the Chinese were fascinated with them," says Constantin Wild.

According to the gem merchant, balance, order and love of structure are at the heart of French high jewellery, which has defined the field of expertise for centuries. "The French are excellent at mixing symmetry and refinement, slightly frivolous in the best possible way, but also with the extraordinary elegance of French formality, when they so desire," says Constantin Wild. "This really sums it up: gems are international good taste. Always stylish, chic and surrounded by the imperceptible scent of money. Gems appeal to the whole gamut of people's emotions. They are the common language of the worldly, communing with the extremely discreet – but they can also be the non-verbal communicative language of the competitive and ostentatious fashion business," says Constantin Wild.

"That a gemstone enchants the eye is its very raison d'être. Gems sparkle with history, ancient tradition and contemporary pleasure."

Constantin Wild

FROM SCULPTURES TO SAND CLOCKS

Constantin Wild had a long-standing and very warm friendship with the artist Manuel Carrera of the Spanish jewellery company Carrera y Carrera. Manuel Carrera would ask Constantin Wild to source some of the most unusual, remarkable and exciting gems, as these clearly helped him develop and define his style. Some of the most popular pieces in Carrera's bespoke production were solid gold figurines clad in gems. Just such a piece was the sculpture of a Moor with an opal, over which Constantin Wild and Manuel Carrera formed a bond of friendship. "After that, we did many more interesting jewellery and precious art projects together. Manuel was one of my first customers, and he ordered a lot of exciting things, among them figurines carved out of rock crystal and lapis lazuli, stone bases cut of opal, malachite and jasper for his art works, and more," Constantin Wild tells the story. Manuel Carrera would involve Constantin Wild in the production of his art pieces, and together they realised some of his most outstanding creations: champagne glasses cut from amethyst and rock crystal, agate columns for large medieval-styled silver boxes, and exotic, fun projects like the gem parts for a large hourglass for an Arab client's pleasure house.

A HISTORICAL OVERVIEW

Since **the second half of the 16th century**, colour has been paramount in jewellery, either as enamel or in gemstones. In the 19th century, jewellery became more than just a mark of status; the design of a jewel became a matter of interest in itself.

During **the second half of the 19th century**, 'revival' jewellery was based on unearthed antique artefacts. In addition to cameos cut from layered stones, naturalistically depicted floral jewellery was made with enamelling.

Art nouveau (1895-1910) was characterised by the use of enamel, with René Lalique as one of the most important figures, pastels and greys (moonstones for example), free-flowing curves, asymmetrical, organic motifs, the use of horn and ivory, and Japonism. **Edwardianism** (1901-1910) played on diamond-set elegance and the graceful and classic garland style; favourite motifs were lilies, ferns and laurels, in an all-white look with pearls.

The **Roaring Twenties** took over with exoticism (African tribal pieces and Chinese, Japanese, Middle-Eastern and Egyptian inspiration), modernism and art deco and a vibrant palette of colours garnered from the use of emerald, rubies, sapphires, turquoise, tourmaline, onyx, aquamarine, carnelian, coral and amethyst. The excavation of Tutankhamen's tomb in 1922 inspired col-ourful Egyptian motifs in jewellery – just as the Ballets Russes in Paris engendered a lust for primary colours after its première in 1909.

The 1925 exhibition 'Exposition Internationale des Arts Décoratifs et Industriels Modernes' lent its name to the art deco movement, confirming the streamlining of looks; more 'culture, machines and straight lines' in pure geometric forms, finally culminating in an all-white look again.

The 1930s were more sober, as the decade was characterised by economic depression and political instability, but those affluent enough to enjoy the high life of café society did so with the glamour of a bejewelled existence. The 1940s saw more three-dimensional floral jewellery once again, returning to naturalism. Splashes of colour and a certain tongue-in-cheek effect were achieved with the cocktail style. Large colourful brooches and bulky glamorous pieces set in gold (or at least gilded) with colourful gemstones (or paste), humoristic motifs and large abstract rings set with citrines and aquamarines lifted people's spirits during two difficult decades.

The 1950s and 1960s enjoyed a glamorous jet-set style, but the economic boom was followed by controversy. The 1950s jewellery scene saw matching sets with whimsical motifs, and Bulgari began setting cabochon-cut coloured gems en masse as prominent elements in the company's jewellery, using their combinations to enrich visual impact. When pop art entered fine art, jewellery became design-driven with small diamonds set in yellow gold. The 1960s enjoyed fruits and vegetables as motifs in jewellery, which Jean Schlumberger (1907-1987), for example, interpreted for Tiffany & Co.

Since **the 1970s**, individual 'salon designers' of high jewellery have made their presence felt internationally; among these are David Webb, Oscar Heyman, Wallace Chan, Cindy Chao, Anna Hu, Michelle Ong, JAR, Hemmerle and Otto Jakob.

The pages that follow are Constantin Wild's celebration of exceptional creations from some of the world's most iconic jewellers. Set in jewellery, the gems have taken a privileged part in a stylistic dialogue with the high jewellery designers, goldsmiths and gem setters of some amazing companies, capturing joy, power and love. Just as cutting brings out the hidden beauty of gems, jewellery and design ideas make the most of their fabulous colours and intoxicating rarity. Firmly founded in history and extraordinary gems, today's jewellery industry expresses and embraces the creativity inherent in them. Some Constantin Wild gems have been used historically in the jewels of past centuries, and naturally they continue to show their true colours in the jewels manufactured today.

Gems set in high jewellery are akin to subliminal promises that touch the heart and soul of the wearer. Like a lover's poem, jewellery is all about confident elegance – creating unique emotions with impeccable integrity and profound value.

Necklace and earrings from the 2019 Blue Book Collection, Jewel Box Set made of 18-carat yellow gold set with imperial topaz - the necklace with over 75 carats and more than 11 carats of diamonds - the earrings with over 23 carats of imperial topaz and brilliant-cut diamonds.

Tiffany & Co.

Tiffany & Co.'s high jewellery is a study in virtuosity which magnifies and illuminates Nature's beauty with colourful intensity, showcasing the earth's most unusual and exquisite gemstones.

In 1837, Charles Lewis Tiffany (1812-1902) founded his company in New York City, and soon the store was acclaimed as the palace of jewels for its exceptional gemstones. Since then, the company's love affair with coloured gems has been varied indeed, with proponents such as chief gemmologist George Frederick Kunz (1856-1932), in-house designer Jean Schlumberger (1907-1987) and artist Paloma Picasso (born 1949).

The famous Tiffany Blue Book is the brand's annual showcase of unparalleled works of high jewellery set with superlative gems that push back the boundaries of design. Since its first issue in 1845, the book has been the ultimate in high jewellery design. These spectacular one-of-a-kind creations are handcrafted by artisans in the workshop above the Fifth Avenue flagship store, in whose hands Tiffany's legacy comes brilliantly to life. The high jewellery speaks to connoisseurs of the unusual and unexpected, to people who are looking for something they have never seen or worn before. The Blue Book also highlights Tiffany & Co.'s gemstone heritage, especially the deep purplish-blue of tanzanites, the brilliant green of tsavorites and the pinks and purples of kunzite and morganite – gems either discovered and or named and marketed by Tiffany & Co. as the first jewellery company to do so. Colour-saturated jewels with these four gems, plus many other vibrant gemstones, often line the pages of the Blue Book.

Charles's son Louis Comfort Tiffany was himself described as a 'brilliant col-ourist'. In 1879 he hired the exceptionally knowledgeable and enthusiastic gemmologist George Frederick Kunz, who brought an unparalleled variety of minerals to Tiffany & Co.'s jewellery during the late 19th century. Kunz travelled the world in search of unusual gemstones, and his writings established his own name internationally – and Tiffany's reputation in coloured gemstone jewellery for ever. During his tenure at the company, Kunz introduced the firm to deep-purple Siberian amethysts, brilliant Mexican fire opals and dark green and pink Maine tourmalines. Kunz also assembled specimen collections for museums and private individuals, two of which were purchased by John Pierpont Morgan (1837-1913) for the American Museum of Natural History in New York. The pink beryl was named morganite after Morgan, and kunzite, a pink-purplish spodumene variety, was named after the chief gemmologist himself.

The Blue Book

The bejewelled birds, flowers and ocean life of the legendary, elegant and witty Jean Schlumberger have been regular features of the Blue Book since 1956. After World War II, Schlumberger came to represent and reaffirm Tiffany & Co.'s international acclaim. Trained as a designer in France and employed by Elsa Schiaparelli (1890-1973) during the 1930s, Jean Schlumberger brought a large-scale whimsical aesthetic of fashion jewellery to the world of fine and high jewellery; he had large, glamorous and colourful stones set in swirling gold. He was inspired by Nature, but treated it as an inspiration for fantasy, producing animal and plant forms in jewellery that was baroque in exuberance and modern in line and form. His love seemed to permeate his every piece, and Tiffany & Co. once wrote with good reason that their "rarest treasure is a genius called Schlumberger". Schlumberger treated serious gemstones with a light touch, and his verve and energy had American clients clamouring to adorn themselves with his creations; his originality was universally acclaimed – and adored. Schlumberger flamboyantly used coloured stones in massive sizes, and his work was generous, quirky and humorous, attracting a celebrity clientele.

Winter bracelet from the 2018 Blue Book Collection, 'The Four Seasons' by Tiffany made of platinum with mixed-cut aquamarines - over 26 carats in total - and mixed-cut diamonds - over 26 carats in total.

'Bird on a Rock', designed by Jean Schlumberger

Jean Schlumberger had a high-profile reputation, the best social contacts and an indisputable talent for colourful jewellery design. He was dubbed the "most creative and influential fine jewellery designer of the twentieth century", owing to the wild artistry of his creations that displayed a sense of freedom and fantasy. Schlumberger's inspired use of gemstones was like a full blast of Technicolor, almost an explosion of colours in jewellery.

Jean Schlumberger

Melvyn Kirtley, Tiffany & Co.'s chief gemmologist emeritus after 30 years' tenure, has said the following: "Schlumberger used an incredible palette of gemstones, and he loved all gemstone types. Depending on the colours he needed, he used them all as his palette. Schlumberger wanted gems clear without inclusions, but primarily, he looked for saturations to fit his designs. He used varied cuts, but mostly ovals and cushions. 'Bird on a Rock' was unique for Schlumberger's artistic production, because the gem had to be for a certain size. We have three different bird sizes – small, medium and large – to fit different-sized gemstones and their orientations. It is a painstaking process to match the depth and ethereal qualities of the jewel-like tones that he always strove for. He looked at Nature, saw its beauty and recreated it in a surreal way that had fantasy about it. That made him stand out from his contemporaries. He would have loved paraíbas for their colours and the way he could have worked with them; they would have stimulated his imagination."

Paloma Picasso

In the 1980s, Paloma Picasso's designs captured the informal fun of the decade in seminal 'power jewels'. When John Loring (born 1939) took over as Tiffany's design director in 1979 (today emeritus), he – as a long-standing friend and mentor to Picasso – encouraged her to try her hand at designing jewellery. She had trained as a theatrical costumier and stylist in

Paloma Picasso by Nigel Perry

Designed by Paloma Picasso for the 1985 Blue Book, this necklace of rare and colourful gemstones was acquired in honour of her 35th anniversary with Tiffany & Co.

Necklace from the 2019 Blue Book Collection, Jewel Box.
Made of platinum and set with pear-shaped and oval cabochon moonstones - over 121 carats total - and more than 14 carats of diamonds.

Paris, improvising stage jewellery and designing costume jewellery for Yves Saint-Laurent (1936-2008). Gold was always Picasso's primary metal and material, allowing her to focus on the gemstones accumulated through decades by Tiffany & Co. In the 1980s, loud colours were worn with confidence, which made it the perfect moment for her to step on to the Tiffany stage. Her love tended towards big gems in strong colours – rubellites, tsavorites, tanzanites, bright pink and green tourmalines, kunzites, sky-blue aquamarines, citrines and orange Mexican fire opals – for which she designed abstract, strong pieces. Her jewels intermingled with the 'dress for success' idiom of the 1980s in a bold and joyously graphic style which clearly delighted in large gems. Paloma Picasso's use of assertive colours excels in dazzling kaleidoscopic combinations, in a style described as "high-impact thanks to deep colour mixes". Sleek and tailored, Picasso's jewellery designs favour distinct faceted and unusually shaped gemstones which give character to her creations.

A passion for beauty

As for the long history of Tiffany & Co.'s work with fantastic coloured gemstones in high jewellery, the Tiffany Blue Book Collection has shown the company at its most creative, experimental and innovative. The collection is connected to icons and language that are part of the history of Tiffany & Co., but reimagined in new ways – manifesting the company's tradition for sourcing some of the world's most rare diamonds and coloured gemstones of unsurpassed quality. High jewellery is really meant to be worn beyond a formal setting. It is meant to become part of someone's life. While there are a number of pieces that are created to be worn at a particular moment in time, the vast majority of pieces in the collection can be worn multiple times in multiple ways and really become a part of the wearer's wardrobe and his or her own personal style, from the simplest to the most extraordinary. The Blue Book Collections embody Tiffany's passion for elevating the beauty of Nature and reinterpreting it through the unparalleled craftsmanship that has defined the jewellery house since 1837. Inspiration is drawn from Nature but combined with elements of wit and surprise as opportunities to bring incredible beauty to important clients who are also collectors. The settings that hold these incredible stones were constructed with great care and technical precision, so that the gem's beauty is perfectly framed.

Necklace from the 2021 Blue Book Collection, Colors of Nature. Double-strand convertible necklace in 18-carat yellow gold with 15 round Russian demantoid garnets of over 22 carats in total and mixed cut diamonds of over 34 carats in total.

David Webb

With intricately carved coloured gems making up the larger part of grand-scale necklaces and bangles, David Webb's (1925-1975) eponymous company has pushed back the boundaries of the jewellery imagination for decades – and as a client once said: "To wear David Webb, you must have courage." Truer words have never been spoken.

For more than 70 years, David Webb has been the quintessential American jeweler of highly original, modern jewellery. Best known for distinctive carved and enamelled animal bracelets, dramatic gold necklaces, colourful sautoirs, Maltese cross brooches and exquisitely carved pieces, David Webb is one of the country's most important, distinguished jewellers. All the jewellery is made on the premises in New York, as it has been since 1948.

Creativity was always David Webb's hallmark, and he developed it while an apprentice to an uncle who was in jewellery manufacturing. In 1948, he made it to New York, opening his own shop. By the end of the 1960s, the smart set, the fashionable set and all of Hollywood were wearing their David Webb animal bracelets. David Webb's beloved yellow gold was used in sculptural, fluted and scalloped show-stopping pieces of jewellery. Rock crystal bracelets, carved lapis pendants, totem pieces of brightly coloured gemstones in the most unexpected but perfectly proportioned combinations – jewellery with bold geometric motifs.

In 2009, the present ownership of the company passed to co-owners Mark Emanuel and Robert Sadian, who – drawing on an archive of nearly 40,000 designs – have faithfully restored David Webb's original ones: colour is bright, but never loud – black, white, red, green, off-white, brown, cobalt blue, turquoise blue and on occasion purple. The New York Post wrote: "Mixing stones and multicolour is a Webb signature." For Webb, big was often better, particularly when it came to coloured gems. He was drawn to textures and carved surfaces on stones, and to accommodate uniquely shaped (fluted or carved) gems, Webb designed a number of 'frames' – simply adopted designs – to contain the actual gems he had on hand for brooches.

David Webb's jewellery style is uniquely American with an 'anything goes' optimism; forthright, fresh, elegant, stylish and beautifully made. The colours are striking, volumes generally large, the look visually strong – and the results? Always modern, as the company's catchphrase goes.

Mikimoto

Since 1893, Mikimoto has celebrated the natural beauty, exceptional quality and timeless elegance that define the originator and world's foremost purveyor of the finest cultured pearls. In Kokichi Mikimoto's pioneering spirit, spectacular pearl collections continue to pay homage to natural origins and evolve with high jewellery, a category that is seeing an increasing number of precious gems. With exquisite craftsmanship, fantastic gemstones appear throughout the collections, dancing in festivals of colour. Inspired by Nature and the spirit of the elements, white South Sea and Akoya cultured pearls are combined with gems, paralleling the femininity and strength of their wearer. Mikimoto's high jewellery remains a beautiful marriage of tradition and innovation, embodying creativity, high-grade design and outstanding craftsmanship.

Marina B

When New-York-based, French-bred, Italian-born Guy Bedarida took over as creative director and designer for Marina B in 2017, it was his life-long dream come true. The legendary Italian 1980s brand, founded in 1976 by Marina Bulgari, has proven to be a heritage darling of the jet set. "When I met Marina Bulgari, I was really impressed by this tall, elegant and determined woman, who at the age of 88 still possessed the sharpest eye and a unique point of view. She was the queen of colours," Guy Bedarida has been quoted as saying. Driven by a sense of adventure and ambition, Marina Bulgari left her family's legendary house in the 1970s in order to launch her own brand. Her trailblazing designs were bold, timeless and contemporary, and within a decade, she had clients the world over. Irreverence to colour codes, choices of stones and material mixes made Marina Bulgari's creations rich and opulent. Her reputation quickly grew as solid as the stones she joyously laid on rings, bracelets and necklaces. Marina B's modern vision and creative excellence inspired the first generation of women who bought jewellery for themselves.

"My first inspiration is the archive of Marina B – bold, opulent and colourful creations; my second source of inspiration is the gems that I search for tirelessly. When I travel, I am always on the lookout for new, rare and unusual coloured gems."
– GUY BEDARIDA, creative director of Marina B

Chopard
Red Carpet Collection

Chopard

Chopard, the artisan of emotion, displays a unique use of gems that is unfailingly inventive and bold. Creativity, passion and exceptional artisanship are the key to the *haute joaillerie* creations that feature many traditional shapes – square, round and oval – as well as original ones and a large range of colours, for a very diverse and imaginative look. Here, Chopard's co-president and creative director Caroline Scheufele speaks about the company's use of coloured gems, all set in 100 per cent ethically sourced gold and platinum.

"All the coloured stones have a story to tell. If you think about how long they have been beneath the Earth's crust and the extraordinary way they are produced – under such high pressures and temperatures for more than a billion years – it makes the dinosaurs look young. Emerald is one of my favourite stones, as it represents the energy patterns of the activated heart chakra, the fountain of the emotions; rubies are another love of mine. Most of the time I enjoy mixing coloured gemstones with diamonds in order to magnify all their colours. I like to create jewellery pieces which combine gemstones with unexpected elements, for example an amazing and flamboyant feather necklace, inspired by the traditional costumes and crafts of the Mongolian plains. Representing a major first for the maison, this creation combined a magnificent centrepiece – composed of a sculpted gold cameo adorned with swirls of blue apatites, violet garnets and red jasper – with a collar made of rooster, grey heron, ostrich and black pheasant feathers, designed and realised in cooperation with feather artist Nelly Saunier. For the Red Carpet Collection 2019 and for the first time, I also combined coloured gemstones with wood."

"Chopard is always looking for innovation, from design to production; traditional craftsmanship always meets cutting-edge technologies. For that reason we have been using new materials such as titanium. We used the titanium to enhance the colours of our gemstones, but also to give lightness to our creations and make them comfortable to wear. The *La Parisienne* set and the orchid earrings were partly made out of titanium, a metal that I appreciate because of its suppleness and technical nature."

"Love, often the theme of both our fine and high jewellery, is an anthem to coloured gemstones at Chopard. Throughout the years, the Red Carpet Collection has also been inspired by many other themes: the animal world, Nature, art history and of course all the beautiful actresses who wear the creations at the Cannes film festival, the Oscars and many other award shows."
"When I design high-jewellery pieces, one of my priorities is to make sure I vivify the colours and the light. I have used many tanzanites, as I adore the reflections of this stone, which can vary from blue to purple. The origin of those newly discovered gemstones is also very important for Chopard. In 2018, we created an extraordinary necklace featuring a 34.63-carat triangle-cut paraíba tourmaline from Mozambique, the first responsibly sourced paraíba for this signature collection set with diamonds certified by the Responsible Jewellery Council and 18-carat Fairmined gold. It was unveiled on the famous red carpet of the Cannes Film Festival – worn by Julianne Moore, with whom I collaborated with in designing this special piece. She has been a great friend of the house and supporter of our journey to sustainable luxury for many years."

– CAROLINE SCHEUFELE

Chopard
'La Parisienne'
High Jewellery Collection

Chopard
Red Carpet Collection

Anna Hu

Although Anna Hu may have been active under her own name for less than two decades in the high-jewellery genre, the Taiwanese-born New-York-based daughter of a gem-stone dealer and a pearl and jade expert has long since made global headlines for herself owing to her style, which can best be described as a gem set *in extrema*. Alternating between large, medium and small-sized gems, the client may expect to find every surface of Anna Hu's *haute joaillerie* creations covered in colours, sometimes in gradation – sometimes opposite one large gemstone taking centre stage. And yes, the largest Anna Hu pieces are set with several thousand gemstones; the results are always breathtaking and feminine. Before founding her own company in 2007, Anna Hu worked for Harry Winston, Van Cleef & Arpels and Christie's. Her own design marries Eastern motifs with Western art in extravagant creations which have set auction records for contemporary jewellery. She has had solo exhibitions at the State Historical Museum in Moscow, Paris's Place Vendôme and Christie's. In the 21st century, colour blocks have been given a new lease on life thanks to Anna Hu.

Anna Hu was originally trained as a cellist, and her close relationship with classical music is reflected in her grand jewellery works. Hu has said that she frequently turns to classical music to get her creative juices flowing, and that her process of creating jewellery is better described as "composing". She elaborates on that: "Each note is equivalent to a gemstone, the melody is the curve of the line, the harmony is the combination of colours and the rhythm is the technology in the details."

– ANNA HU

Butterfly Dream
Necklace in kunzite

Siren's Aria Collection
in aquamarine

Suwa

When Suwa's founder Kinomatsu Suwa first encountered a gemstone while visiting his older brother, a metalsmith in Yokohama, his fascination with its beauty was the starting signal for his choice of career. Foregoing his family's traditional business of selling Japanese prints, he went out on his own in 1908, dealing in gemstones. During the 1950s, second-generation Kikuo Suwa came across an imperial jade (jadeite) possessing an extraordinary level of transparency.

He sold it, but after some circulation eventually it returned to Kikuo's hands. Feeling that an item of such rare beauty should be preserved for posterity, Kikuo donated this particular piece of imperial jade, nicknamed 'The Green Pepper', to the National Museum of Nature and Science in Ueno, Tokyo. In 1965, third-generation Yasukazu Suwa became the first Japanese national to earn the degree of Graduate Gemologist (GG) from the Gemological Institute of America (GIA). Since then, Yasukazu has flown round the world in search of beautiful gemstone material for the lovely jewels of his family business, and thanks to his impressive gemmological authorship – three large volumes about gemstones, their quality, value, beauty and appreciation, as well as a dictionary of gemstone specimens, richly illustrated with images of the Constantin Wild inventory – he has expanded Japanese knowledge of the beauty of gemstones.

Pomellato

Synonymous with creativity and colour, Pomellato was founded in Milan in 1967 by Pino Rabolini – the heir to his family's goldsmith legacy. While shaping Pomellato with a powerful, fashion-forward identity, the company's sophisticated gem combinations feature special cuts, rounded or irregular, set in visionary designs, with a modern yet unconventional beauty, immediately recognisable. For Pomellato, the extraordinary gem colours reflect a truly unique and changeable style, inspired by fashion and feeling. The powerful allure of colour-blocking is pure Italian glamour, and the creations enchant the eye with an ocean of brilliant nuances. Since 2015, creative director Vincenzo Castaldo has realised several projects with Constantin Wild, benefiting from the unique and almost unlimited resources of the company, with several unusual gemstones cut specifically for the 50th anniversary line entitled the Ritratto Jubilee Collection.

Bulgari

Cinemagia High Jewellery Collection

Bulgari

In a celebration of life and beauty, the Roman company par excellence – Bulgari – has a bold and brave approach to high jewellery. Vibrant colours, fanciful shapes and exquisite details convey a sense of joy and positivity. This iconic aesthetic, defined by sinuous lines, voluptuous shapes, intricate complexity and unpredictable fascination, is revisited season after season, seen through Bulgari's contemporary and experimental filter for sumptuous, versatile and multifaceted creations, rewriting the rules of high jewellery design and manufacturing.

With the clear ambition of bringing exuberance into the world, Bulgari has inspired awe and wonder through the use of gems since the 1950s. Upscale preciousness is matched with effortless elegance. Rare, exclusive gems are found around the globe by Bulgari's creative jewellery director Lucia Silvestri and forged by the brand's skilled artisans at the workshops in Rome. Harmony, grace and beauty stand out in Bulgari's one-of-a-kind pieces, with their vivacious hues of a powerful kaleidoscope, cementing the company's unlimited commitment to beauty. Whether cabochon or faceted, this creative bravery is a triumph for *joie de vivre* or, as the Italians would say, *la dolce vita*. The high jewellery is designed to steal the spotlight on red carpets around the world, and an outstanding colour palette of charming, delicate or vivid shades of the rainbow, colour combinations, gemstone cuts and artistry comes together in highly coveted masterpieces.

Since the 1940s, Bulgari has found the company's ideals in Greek classicism and Roman antiquity, to which a mix of Italian renaissance style and power colours has been added. The virtuoso use of colours is a sure signature of high-jewellery Bulgari creations. Since the 1970s, Bulgari has gravitated toward abstract motifs, and starting in the 1980s, the company has flaunted superior gold work. Thanks to the gorgeous gems and the use of this most precious of metals, Bulgari's jewels have a particular sense of volume. The visual expression is all about strength communicated through colours. A desire for conspicuousness continues to sum up Bulgari's colour attitude in high jewellery that possesses a life of its own.

Bulgari

Cinemagia High Jewellery Collection

Bulgari

Cinemagia High Jewellery Collection

Temple St. Clair

When Temple St. Clair founded her eponymous company in 1986 in Florence, Italy, beginning her collaboration with the world's finest goldsmiths, her brand was the first fine jeweller to be represented and carried by Barneys New York. With an artist's eye and an explorer's heart, St. Clair fashions rare coloured gems with distinctive gold work to illustrate universal narratives of the Earth and the cosmos. Her high jewellery collection – representing artistry at its most collectible – celebrates a connoisseur level of gemstones with one-of-a-kind pieces born of an ongoing quest for the rarest stones in the world. These include richly patterned Lightning Ridge Australian black opals, electric blue Brazilian paraíba tourmalines, fiery African tsavorites, royal blue moonstones, purple, demantoid and colour-changing garnets, and diamonds in numerous nuances. Temple St. Clair holds a Master of Arts degree in Italian Renaissance Literature from Middlebury College; in 2016, she received the prestigious GEM Award for Jewellery Design for her work on the *haute joaillerie* collection The Golden Menagerie. In 2017, Temple St. Clair joined Louis Comfort Tiffany and Alexander Calder as the third American jewellery designer whose work is represented in the permanent collection of the Museum of Decorative Arts at the Louvre in Paris.

"Basically, I am a colourist at heart. I came to jewellery through a love of art history and literature, and I use gemstones in a painterly way; I am a perfectionist and colour is key. I tell universal stories through jewellery design, and I go to Constantin Wild for the best material in the world – the best colours and the finest quality."

– TEMPLE ST. CLAIR

Theo Fennell

Original. Beautiful. Perfectly crafted in order to give pleasure for generations. Theo Fennell believes that the genius of jewellery (and silverware) is in the details, in caring as much about the back as the front. This attitude shines through in the final piece. In his unique works over more than 40 years, Theo Fennell has demonstrated that he is not only a disciple of the great traditions of British jewellery and silver, but also at the vanguard of new techniques and materials. Although Theo Fennell's upbringing exposed him to many cultures, his designs have that quintessentially British sensibility; music halls and musicals, churches, pop art and music, theatre, literature, sport and architecture, fairgrounds and Greek classicism, Elizabethan art and poetry are just a few of his sources of inspiration. The jewellery – however simple or theatrical – is always unique and beautifully made. Patrons all over the world love and understand Theo Fennell's commitment to making superb works for them, adhering to their sense of style and appreciation for real originality and quality. Glorious jewellery for marvellous people. Extraordinary stones are set in fabulous cufflinks in a wonderful combination of materials as miniature works of art.

"I believe that jewellery should be personal and unusual, beautifully designed and crafted, and it should become something that evokes great sentimental attachment. Most of all, good jewellery should always bring real joy to its wearer."

– THEO FENNELL

Lester Lampert

The story of Lester Lampert began in 1904, when his grandfather David Lampert made the brave and arduous journey from Russia to the United States in search of a better life. In Chicago he established what is now a jewellery business spanning four generations of the Lampert family. When the time came for Chicago's perhaps best known gemstone museum, Chicago's Field Museum's Grainger Hall of Gems, to choose a company to entrust with their most noteworthy gems for setting in jewellery, there was no doubt in the mind of senior vice-president Lance Grande. Head of collection and research, curator of the Grainger Hall of Gems, Grande said, "Lampert were chosen to create new jewellery articles because of their demonstrated design vision, manufacturing expertise and great enthusiasm for the project. Many of the Field's most noteworthy cut gems were put into Lampert's hands, including the 97.45-carat red topaz, the 153.79-carat peridot and a double-sided 37.90-carat star sapphire." The result was 27 fabulous pieces.

Stanislav Drokin

In the 1980s, Ukrainian-born Stanislav Drokin began his professional life by training as an engraver – and soon realised that the profession is closely intertwined with the craft and mastery of watchmakers and jewellers. He began studying jewellery in his limited spare time, and over several years, at intervals, he studied gemmology in the gemmological centres of Ukraine, Poland and Germany. These disciplines, in combination with his graphic design degree, have since formed the basis for his jewellery designs. In any creative work, the ability to use a variety of methods, techniques and skills, along with comprehensive knowledge in related areas such as gemmology, chemistry, metalwork, casting and computer graphics is a good way to achieve fresh, original objectives. "I was fortunate to gain knowledge and learn the secrets of past masters, and I am also pleased to attain new horizons of knowledge and technologies of the future," says Stanislav Drokin. Since 1994, he has been making his own bespoke jewellery pieces – unique creations that are an outcome of a multidisciplinary approach to design. Above all, the award-winning jewellery designer is known for his colourful creations, and he frequently dips into the treasure chamber of Constantin Wild for sparkling coloured gems.

"I love all gemstones without exception, but when I got acquainted with opals, it was like an explosion of emotions for me! Opals have the whole universe hidden in them. With the help of opals, Nature sends us an extraordinary palette of emotions and a mix of colours, showing their harmony and changeable consistency."

– STANISLAV DROKIN

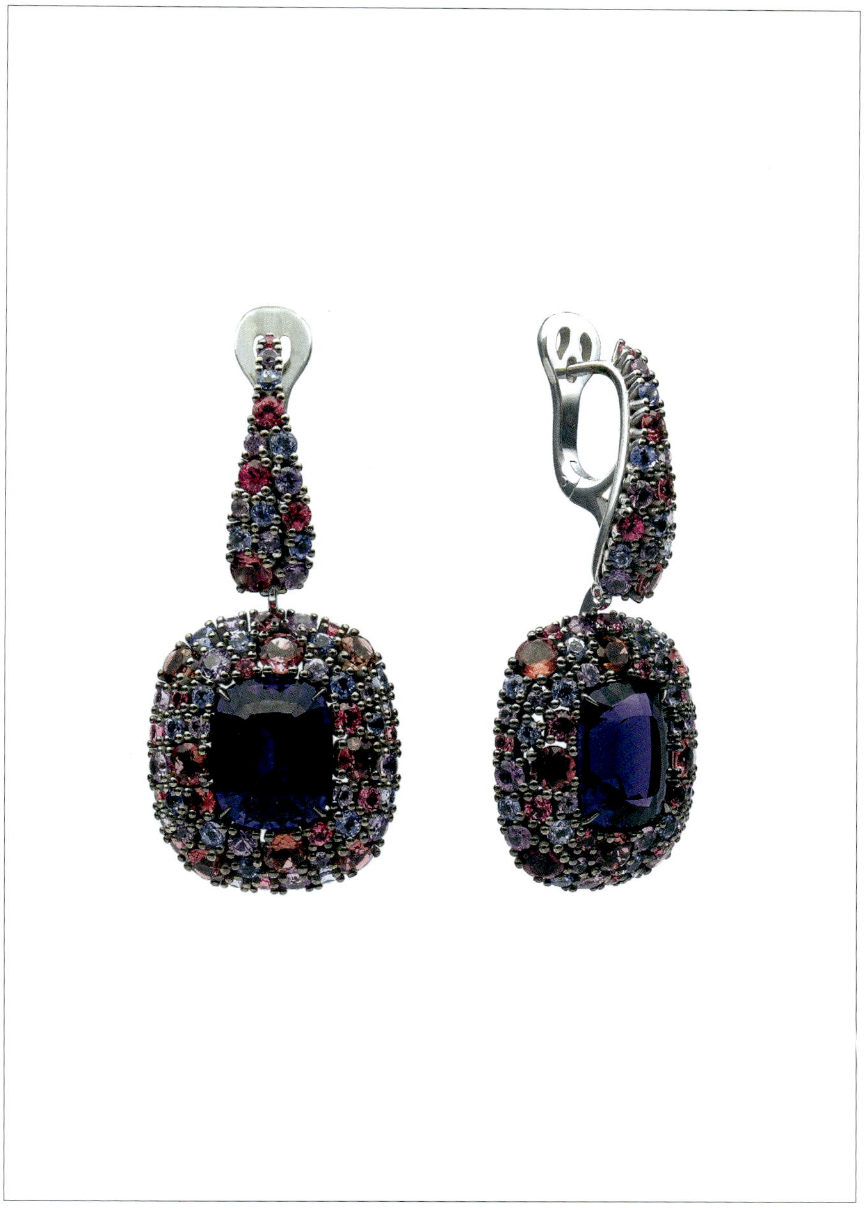

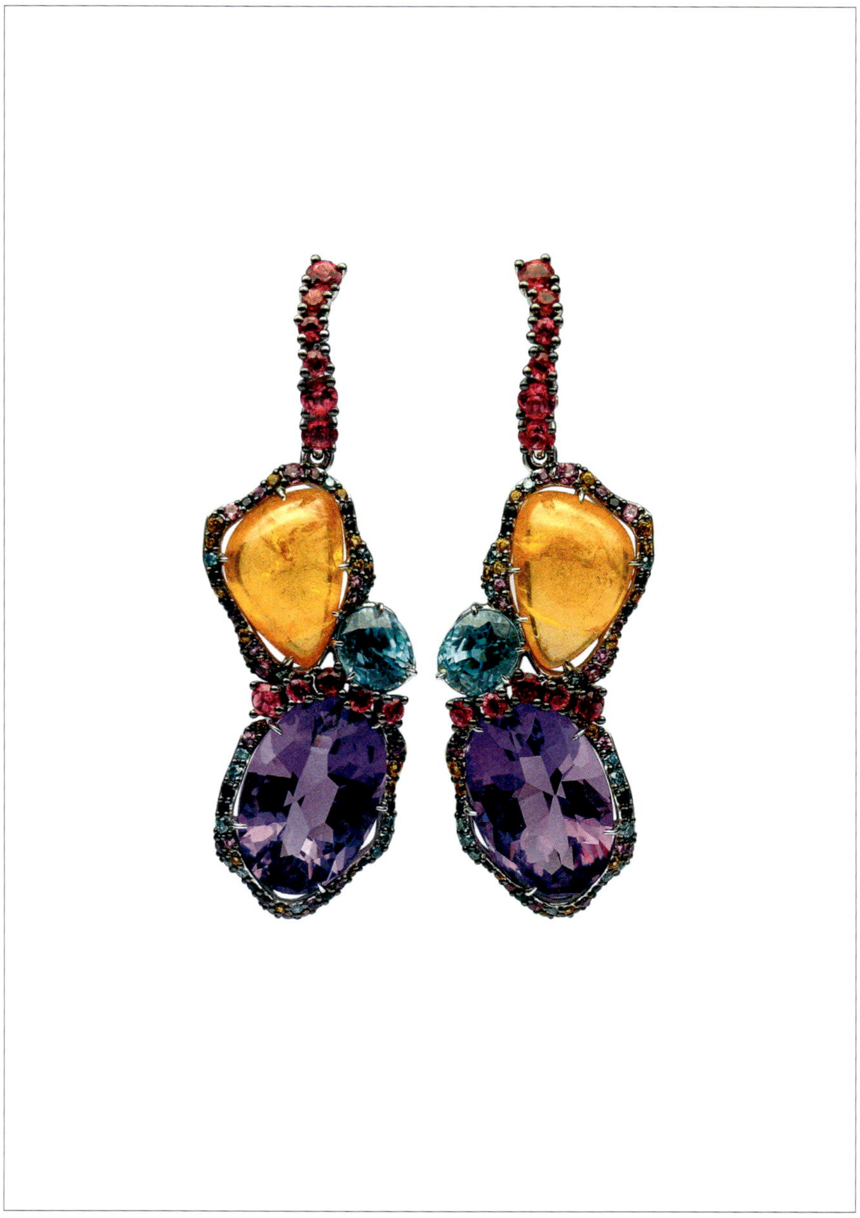

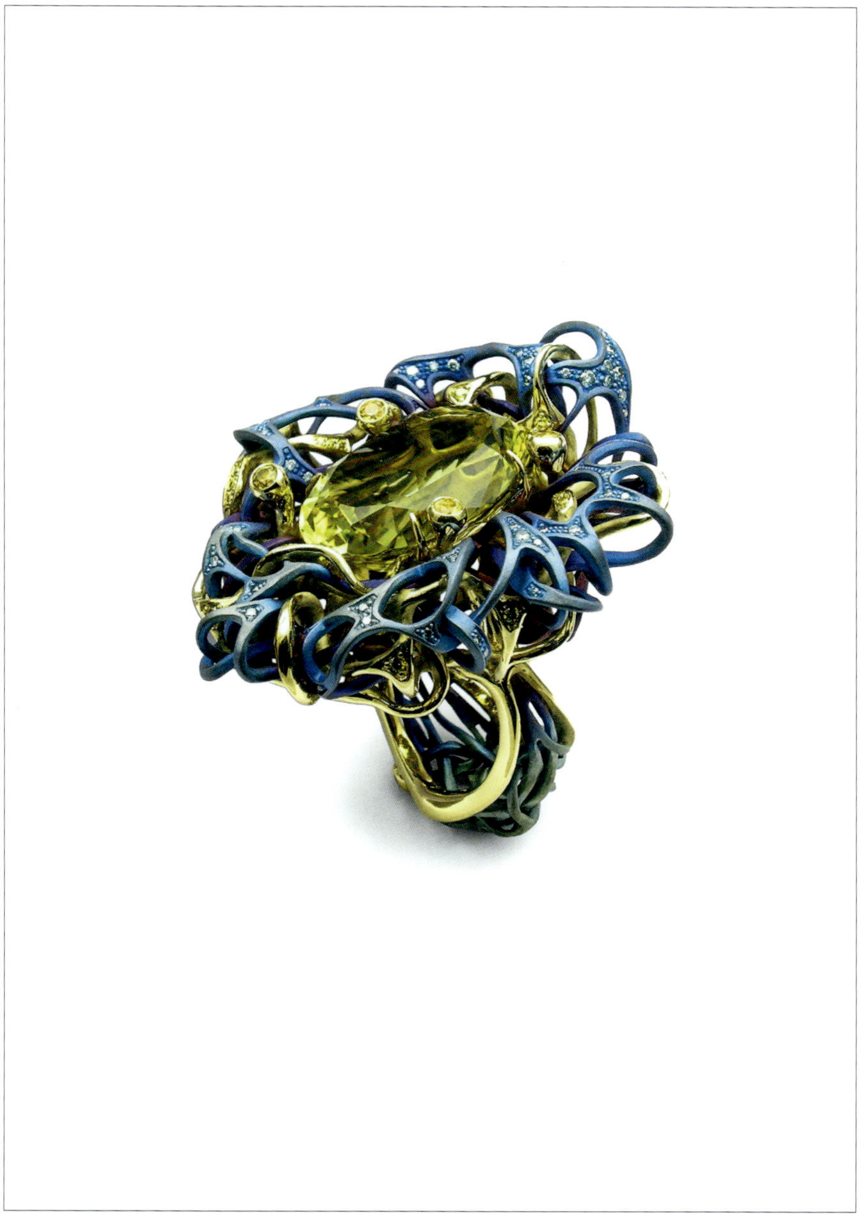

Oscar Heyman

Since 1912, the family-owned and family-run company Oscar Heyman has delighted women with jewellery set with exceptional gemstones. Colour, beauty and rarity stimulate this 'jewellers' jeweller' to craft one-of-a-kind masterpieces, and for more than a century, Oscar Heyman designers, craftsmen and gemmologists have celebrated that process from sketch to workbench. The enthusiasm and artistic vision manifest in the jewellery have been treasured for generations, and vintage pieces signed 'Oscar Heyman' fetch high prices at auctions. Thanks to extensive company archives and a history of exceptional manufacturing, many estate pieces by Oscar Heyman can be authenticated, adding greatly to their allure and value.

At Oscar Heyman's workshop in New York, dreams have been turned into exquisite reality for three generations, and the family remain dedicated to superb gems, elegant designs and impeccable craftsmanship. At the beginning, the company had no storefront; they were simply the discreet workshop where Tiffany & Co., Cartier, Van Cleef & Arpels and others had their fabulous jewellery created, resulting in the trade moniker 'the jewellers' jeweller'. Within a high-style American jewellery tradition, fabulous pieces were made for some of the most elite jewellery houses, establishing a national reputation for outstanding ornaments set with superior gems. Throughout its history, Oscar Heyman became known for setting high-quality sapphires in both traditional and unusual cuts and colours, star sapphires and exceptional and rare specimens such as alexandrites, padparadscha sapphires, tsavorites, mandarin garnets, tanzanites, cat's eye (cymophanes/chrysoberyl) and paraíba tourmalines – all chosen for their beautiful and exotic colours. During the 1970s, the branding of Oscar Heyman's own company name began. The second and third generations of the family continue to pursue high-quality gemstones, using their shapes and colours to suffuse the designs. The jewellery is made under one roof in New York – by lapidaries, gem cutters, metalworkers, CAD workers, toolmakers, gem setters and polishers – and the craftsmen are not only international; several generations of the same family sometimes also work for Oscar Heyman at the same time.

CEO and co-owner Lewis Heyman says, "Oscar Heyman is synonymous with colour, and the gemstones really speak for themselves. Our jewellery contains coloured stone as the centre stone(s), with diamonds serving as accents. An example of our design philosophy is the multicolour gemstone necklace, where assorted sapphires and tsavorites are woven together to form a mosaic masterpiece. Coloured gemstones are at the core of our designs, as we blend the richness of the defined hues with the harmony of individual tones. We constantly search for stones that make our gem buyers stop in their tracks, for instance a suite of perfectly matched paraíba tourmalines, an unheated Burma ruby or a harlequin black opal. Glamour is the excitement one feels in the presence of beauty. Custom-designed projects are among our most meaningful work, as we love the collaborative effort of bringing a customer's dream to life. In 1969, we were asked by Cartier to design and produce the necklace for the 69-carat diamond purchased by Richard Burton for Elizabeth Taylor. The piece is a huge part of our company pride."

"Gems are the centrepieces of the creations. Traditionally, we would mainly work with the big four – diamonds, emeralds, rubies and sapphires – but we are always happy to work with the finest and rarest of stones, which can also be cat's eyes (chrysoberyl), star sapphires, alexandrites, black opals or aquamarines. For Oscar Heyman, high jewellery is recognised not only for the finest qualities of its components, but also for the orchestration of its singular design. Jewellery should transcend time like a fine painting, never losing its appeal," Lewis Heyman concludes.

Chow Tai Fook Jewellers

At Chow Tai Fook Jewellers, the gems are intensely the stars of the show – the diamonds playing supporting roles in the jewellery designs. Founded in 1929, the iconic brand is widely recognised for its trustworthiness and authenticity and renowned for its product design, quality and value. A long-standing commitment to innovation and craftsmanship has contributed to the company's success, together with the long-held core values 'Sincerity • Eternity'. The group's multi-brand strategy comprises the Chow Tai Fook core brand with its offerings of different retail experiences. One of these is Enzo, a coloured gem jewellery brand renowned for its exceptional craftsmanship and contemporary stylish jewellery creations celebrating Nature and emotions, including stars, flowers, oceans, rainbows, birth and love. Embracing the mission of presenting the beauty of coloured gems, Enzo strives to translate its colourful jewellery into delight for customers luxuriating in every moment.

Juwelier Friedrich

Juwelier Friedrich was founded in 1947 by Karl Friedrich (1912-1992) – today located at Neue Rothofstrasse 19 in Frankfurt am Main. Since 2010, the brand has been led by the gemmologist Marc G. Stabernack. Gold and silversmithing is complemented by a joy in design in combination with a knowledge of precious stones and their aesthetic radiance. As a testament to its worldwide customer base, a famous Hollywood diva replied on a live TV show that the best part about Germany for her was: "Jewels from Friedrich". Award-winning treasures made in the Friedrich studio have even reached the court of St. James's, as evidenced by a letter of thanks from the private secretary at Clarence House. The fact that Friedrich produces *haute joaillerie* to match the charisma of the clientele is proven by this stretch-technique bracelet set with colourful gemstones, which Juwelier Friedrich became famous for.

Kinzel & Rall

Since 1842, the German goldsmith and jewellery boutique Kinzel & Rall has been firmly rooted in the heart of the historic town centre of Kempten. From the beginning, the focus was on individual custom-made jewellery, skilfully manufactured at the highest level in their own workshop. In 1987, Hanne and Joachim Rall, both master goldsmiths, took over the company and raised the individuality and craftsmanship of the atelier to a higher level. They expanded the premises with a new showroom and workshop to create more space for new ideas and creativity. The spacious atelier combines traditional artisanship and modern innovative techniques in order to create extraordinary jewellery pieces. With Janine and Ines Rall, the next generation of goldsmiths has joined the company with new ideas and designs influenced by their apprenticeships in Germany, Italy and Norway.

Times, techniques and designs may change, but the essence of Kinzel & Rall remains the same – creating one-of-a-kind fine jewellery for those looking for the extraordinary.

Encased in a detachable golden locket is a Constantin Wild-sourced 10-carat pear-shaped paraíba tourmaline from Mozambique, set in platinum and surrounded by brilliants. A true treasure manufactured in the Kinzel & Rall workshop.

CINDY CHAO The Art Jewel

Trained by a grandfather who was a renowned architect, and a father who was an established sculptor, Cindy Chao's jewelled masterpieces display a profound understanding of the creative process. As her father's first apprentice, she learned various intricate techniques and skills of sculpting, and his words shaped her distinct style: "Life and emotion are the core essences of a sculpture." With the mind of an architect and the hands of a sculptor, Chao established CINDY CHAO The Art Jewel in 2004 with the determination to create every piece of jewellery as a miniature work of art, and as the first contemporary high jewellery artist in 200 years, her creations have been inducted into three of the world's most prestigious cultural institutions: the Smithsonian National Museum of Natural History in Washington D.C., The Musée des Arts Décoratifs in Paris and the Victoria and Albert Museum in London.

Chao's pieces boast an intricate, asymmetric design that is beautiful from all angles. Routinely, she presents breakthroughs in craftsmanship, offering a dimensionality very few jewellery pieces have been able to achieve, while thus demonstrating that she is an artist who creates with a forward-looking mindset. Some of the pieces are larger-than-life, but still realistic – some feature stunning contrasts, but all showcase exquisite craftsmanship intertwined with meaningful life stories and emotions.

From global auction houses to the world's most prestigious art fairs, Chao has set and broken records while winning important art awards worldwide. In addition to setting numerous auction sale records in New York and Hong Kong, CINDY CHAO The Art Jewel has exhibited at fairs such as the Paris Biennale des Antiquaires, Masterpiece London and TEFAF Maastricht.

With the world's rarest gemstones, Chao transforms titanium, one of the strongest metals known to man, into ethereal jewellery sculptures with the most organic silhouettes. Admired by an ever-growing legion of international celebrities and style icons including Julia Roberts, Amy Adams, Anna Wintour, Salma Hayek, Ziyi Zhang and Bingbing Li, CINDY CHAO The Art Jewel has become one of the most sought-after high jewellery brands to grace globally prominent red-carpet moments such as the Oscars, the Cannes Film Festival and the MET Gala.

2018 BLACK LABEL MASTERPIECE XVIII
PEONY BROOCH

2018 BLACK LABEL MASTERPIECE XVII
GREENOVIA BROOCH

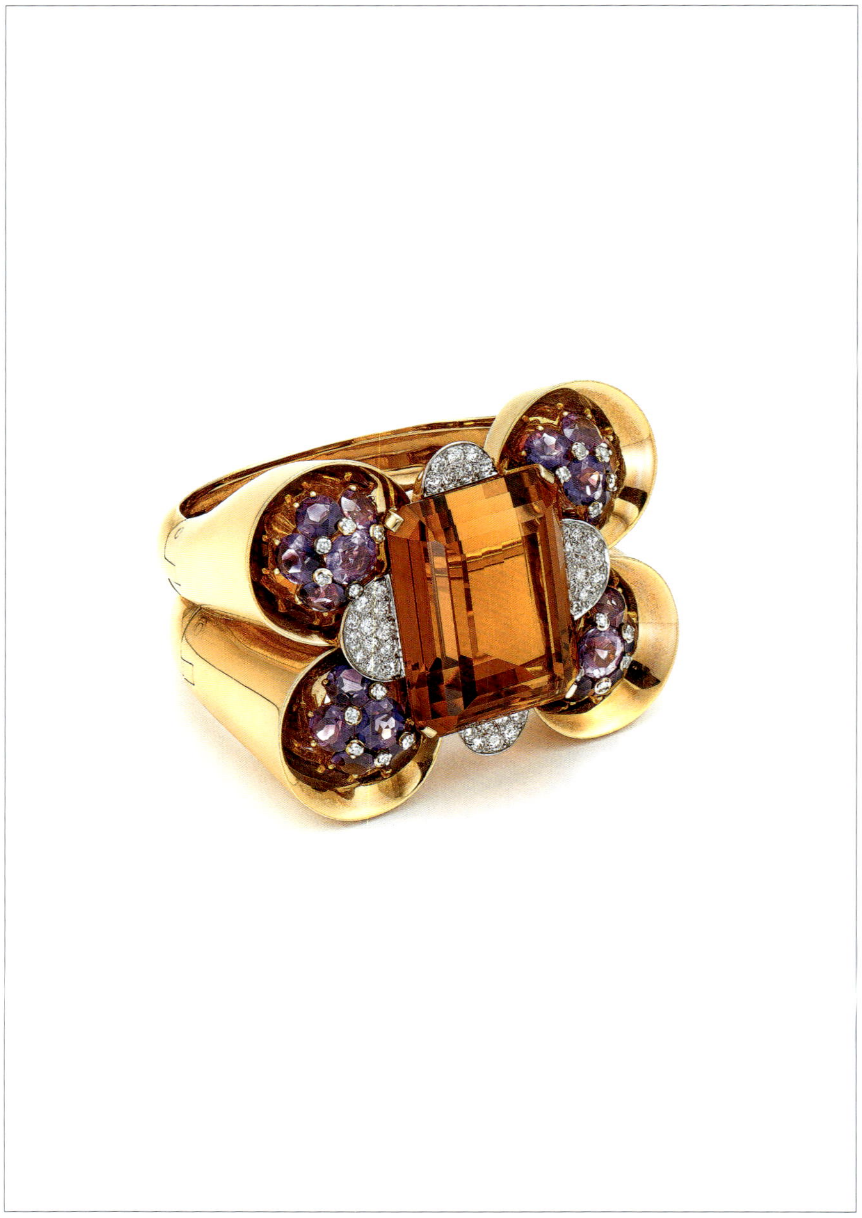

Fred Leighton

For those seeking something unique, a pedigree piece, a design of audacious style, there is only one source: Fred Leighton, curator of the most coveted vintage jewellery. Be it a rare piece from the Georgian, Victorian, Edwardian, art nouveau, art deco or contemporary eras, Fred Leighton has been a trusted authority in the field for more than four decades. From intricate gold designs of the 1800s to bold, sculptural pieces of the 1970s, Fred Leighton's collection inspires clients to wear jewellery in modern ways. Although important period jewellery is the heart and soul of Fred Leighton, the company also creates new pieces that draw inspiration from historical periods. Designed by an in-house team, everything in the collection is signed 'Fred Leighton' and is a limited edition or one-of-a-kind. Here, Fred Leighton CEO Greg Kwiat speaks about coloured gems.

"We curate important pieces from the past 200 years – including many wonderful pieces with wonderful gemstones. We celebrate great style, and coloured gemstones are at the centre of importance for stylish jewellery. Coloured gems form the basis of telling stories with jewellery, and so we are always looking for pieces with important coloured gemstones. After all, beauty is best reflected in self-confidence, and glamour is about self-confidence and a unique style. We look for jewellery that is different and unique, that has a strong sense of style. It does not have to mean today's style, but if a piece is interesting or indicative of a period, these are the creations that excite us the most. Personally, I am a sucker for amethysts and everything purple – I even named my daughter Violet – but because amethysts are not rare, they are not that expensive. However, they are much used in antique jewellery, for example in multiples and/or set with interesting colour combinations. Jewellery with a wonderful, rich depth of colour – that is when the piece sings."

"An important jewellery set for Fred Leighton would be the aquamarine bandeau by Mellerio from the mid-1920s, made in Spain at the time for an important client. It is detachable, so it can be worn in smaller versions as a bracelet or a piece of hair jewellery. The colours of the aquamarines are rich and beautiful – the design very art deco, but it still feels very current. It is incredibly wearable and a wonderful piece. It is a once-in-a-lifetime piece and it stays with you. The Trabert & Hoeffer Mauboussin cuff bracelet from the Reflection series is bold and sizeable, set with a 150-carat citrine; it is a true museum piece."

"Our citrine and amethyst necklaces have a style popularised most recently by VOGUE editor-in-chief Anna Wintour – a layered style of wearing necklaces that makes for a fresh and colourful impact. We like to offer pieces in larger sizes for bigger statements – but of course we also carry them in smaller sizes. Even though we create jewellery at Fred Leighton using older cut diamonds and a few more colours, vintage jewellery is still at the core of the brand. We are known for loaning out vintage pieces to celebrities walking red carpets, to show that the pieces are still modern and exciting, and thereby bringing them back to the spotlight through fashion. Many have collected jewellery for antique reasons, but now, vintage pieces are considered highly fashionable and interesting in themselves."

"I find coloured gems very interesting, and I get to see new and exciting pieces every day, even after some twenty years. There are still stylistic differences between what appeals to Europe, America, and the Asian countries; however, certain style trends have gone global and universal. European sensibility tends to favour slightly smaller, more sophisticated design – Americans usually like bigger pieces. However, that is not true every day or for every person. For Latin America and the Middle East, big is often key – the scale of something makes it feel important. My experience with the Asian market is that they are often looking for something a little more petite. Americans have long been attracted to all shades of blue – red clearly holds appeal for the Asian market. Blue and green still outsell red in the U.S., but colours have universal appeal."

"My grandfather would always say: 'Sell and regret later. But sell first.' That is the fundamental difference between a collector and a person in this business. My grandfather cut a lot of important stones, but he sold them all – as a person in this business should."

Doris Hangartner

Doris Hangartner loves rare, high-quality gems with stunning colours and showcasing their beauty in one-of-a-kind jewellery pieces. Each gem is selected for its superior colour, unique cut, size and even provenance, which means that it may have come from a mine that is no longer active. "I choose to focus on the more recently discovered gems such as paraíba tourmalines with a colour range from Caribbean blue to grass green. Since their discovery in 1987, prices have multiplied enormously," says Doris Hangartner. Her love of and fascination for gems was sparked at a young age by a rock found in the mountains. On the outside, it looked like a plain grey rock. "But the sparkles and shimmer that became visible when it was cracked open left me mesmerised. Since then, I have never ceased to be amazed by the beauty hidden inside rough stones that, when cut, become extraordinary gems." As Doris Hangartner studied, completing her Graduate Gemologist (GG) degree at the Gemological Institute of America (GIA) in Santa Monica, California, she learned more about gems and became passionate about sharing special stones with connoisseurs who appreciate their beauty. "The experience is profound and very physical. Literally, when a match is made, a person's face glows. One person will gasp with delight while another might gaze silently in awe and amazement."

In 2014, Doris Hangartner founded Doris Hangartner AG, specialising in exquisite gems and individualised jewellery design and creations.

PAIR OF RARE GARNETS AND DIAMONDS

Christie's

On 29 May 2018, the auction house Christie's celebrated a triumph at its 'Hong Kong Magnificent Jewels' auction: a one-of-a-kind pair of Russian chrome demantoids – a total of 11.54 carats – were sold for 782,000 U.S. dollars.

From time to time, some of the amazing gems cut and sold by Constantin Wild find their way to the auction houses, where they always fetch solid prices. Whether set in superlative jewellery or kept in boxes, the gems are appreciated by collectors and connoisseurs as the rare natural wonders that they are.

When Constantin Wild had this pair of demantoid garnets in 2017, they were a match made in gemstone heaven. They were close enough in carat weights – 5.93 and 5.61 respectively – to make a stunning pair, and as such they were graded by the AGL American Gemological Laboratories. The verdict? Outstanding in every parameter! The gemstone reports confirmed the Russian origin in the Ural Mountains, no clarity enhancement, excellent colour stability – and that the pair exhibited an intense green colour complemented by a fine cut resulting in numerous bright, internal colour reflections and brilliance.

They also possess an uncharacteristically high, eye-clean clarity, providing a high degree of transparency, and at the time, they represented the best pair of demantoids that the American Gemological Laboratories had ever examined.

When the pair was sold at the Christie's Hong Kong auction, they were set as the centrepieces in a pair of diamond chandelier earrings. When they fetched a price that was even better than the estimate, the world was well and truly alerted to the fact that demantoid garnets of excellent quality make some of the most solid investments – both in financial terms and as objects of wonder.

MEN
WER
SUS-
WO
MEN

Men versus women

Maria Doulton, founder and editor-in-chief of The Jewellery Editor, has spoken to Constantin Wild about the differences between men and women in the way they select gemstones – whether that selection is based on colours, colour changes, asterisms or cat's eyes, or simply on fashion choices.

by Maria Doulton

Colour is the language of Nature. And Nature made men and women very differently. Man the hunter, receptive to the slightest nuances of light and glimmers of movement, woman the gatherer in tune with the chromatic messages in the fruits of the land. Ripe red berries, crisp green leaves or dangerous nightshade hues, woman could read survival in colour, man in shades of light.

PURPLE STAR SAPPHIRE

Animals that we are, men still adopt the camouflage approach with a preference in both dress and gemstones for black, brown, grey and mossy green. Women, attuned to nurturing and preserving the bounty of Nature in all its wondrous forms, respond instinctively to every colour under the rainbow. And nowhere more so than when it comes to gemstones. The miracle of a colour captured for ever in a gemstone is Nature's most powerful voice. Bring on dazzling blue zircon, enigmatic opals, mesmerising pink rhodochrosite, softly glowing moonstone and baby pink morganite – we love them all. There seems to be no limit to the gemstones that catch women's eyes.

EMERALD

CHRYSOBERYL

SAPPHIRE

STAR RUBY

But historically it was men who had first dibs when it came to jewellery. From primitive adornments to crowns, medals and insignia, there was a time when only men were allowed to wear the most precious gemstones. Rubies, emeralds, sapphires, diamonds and pearls were such intense indicators of power that they were donned with clear intentions and strict regulations of rank and hierarchy. Up until the Second World War, a gentleman's wardrobe staples included his gem-set lapel pins, studs and cufflinks. As with their choices in tailoring, men have traditionally favoured stones imbued with sombre authority, aristocracy and masculine decorum.

Today's jewellery lovers will unknowingly be guided by the myriad choices that Constantin Wild makes every day.

"Men tend to opt for more subdued looks and then wear their choices on a daily basis."

Constantin Wild

DEMANTOID
2 carat

PURPLE SAPPHIRE
2 carat

SAPPHIRE
5 carat

COLOUR-CHANGE GARNET
8 carat

At the very start of the journey of a jewel, as a gemstone dealer and cutter, Wild makes ripples through his actions in jewellery shops around the world. Wild has spent his life selecting gemstones, and generations of experience have given him a sixth sense about what will appeal to men or women.

"Traditionally men liked antique engraved stones with seals, such as family crests or warrior heads made from jasper, agate, lapis lazuli, onyx or tiger's eye," says Wild. "Men tend to opt for more subdued looks and then wear their choices on a daily basis."

While the 1980s were all about power dressing – a chunky golden Rolex with a boxy Armani suit – today Constantin has witnessed a move towards men collecting and wearing very precious, often unique and rare gems. They wear them confident in the knowledge that only another connoisseur will recognise the gem's value. Amongst the best-selling are imperial jade, paraíba tourmalines and blue spinels, which particularly appeal to Asian men seeking stones with connotations of power. Special effects like cat's eyes, colour-changing minerals such as alexandrite, rutile needles or star-shaped asterisms are beguiling features that pique men's interest. A good example is Russian demantoid garnet with horsetail inclusions. It is highly sought after and little known outside an elite clutch of collectors. Reversing the usual logic of gemstone valuations, the more feathery the inclusions, the more valuable the stone. One batch was recently sold to a Japanese religious group whose male followers believed in its power and symbolism.

As for global nuances, the American male gravitates towards carved Roman gods and warrior heads in tiger's eye and haematite, while the European man plumps for the traditional signet ring with a family crest in sardonyx, haematite or agate and even precious stones such as sapphires and spinel. But how can Wild divine what will be the next big thing for men? "When I buy gemstones for men," he explains, "I am always led by my own taste. That way if the stone doesn't sell, I console myself by saying that I will make it into a jewel for myself. But funnily enough, after all these years, I don't have many pieces of jewellery. If I see something odd or unusual, with a strange character, and I pick up a loupe and a whole universe of beautiful inclusions comes into focus, the stone has a story to tell, and that appeals to me." Of late, Wild has been intrigued by the cool allure of bluish-grey spinels that, despite his initial hesitation, were snapped up to be made into men's signet rings for the American market.

As for the cuts and settings, there seems to be a gender preference here too. The cleaner light of baguette-cut precious gemstones along with bezel and rub-over settings offers a clean, macho option. Women tend to prefer claw settings and sparklier, show-off cuts with more facets.

With the younger generation, taste has shifted; unisex jewels are the biggest trend in this gender-fluid mood of the moment, and young men are becoming slightly more adventurous in their gemstone-wearing ways. Designers in London report a strong market for bespoke male jewellery, with a tendency to favour whisky-hued chrysoberyls, moody colour-shifting alexandrite, regal red garnet or quirky gemstones such as rutilated quartz. Metrosexual hipsters are turning to designers who offer something new, like a chunky bar of nephrite jade made into a pendant, a green tourmaline ring or a gold bangle set with emeralds and even pearl necklaces as witnessed at the fashion-forward Japanese house Comme des Garçons.

As for women, Wild believes that a celebration of colour is on the cards. "People will want to be happy again, they will party and enjoy life, and so I assume there will be fireworks of colour, blossoming styles with strong joyful tones like tanzanite, rubellite, peridot, green and paraíba tourmalines and garnets, indigolite, yellow citrine, topaz or sapphire, rubies, orange mandarin garnets. Pastels will fade away as we see a second Flower Power on the horizon." And here's a top tip: bright blue zircon is one of Wild's supreme winners thanks to its ratio of colour punch to price.

But just how far will men go? There will always be the peacocks, such as Billy Porter and the late Karl Lagerfeld, who strut outrageous jewellery plumages, delighting us with their choices of outrageous brooches, necklaces and knuckles full of rings. But if mainstream men's fashion is anything to go by, the boys will be sticking to the safer, darker gems and in particular opaque hard stones and those stones that have optical effects.

A little anecdote from Wild captures the spirit of the fine line men tread when wearing jewels. The lining of his formal tuxedo is of bright crimson silk and it matches his 'blood red' rubellite cufflinks from Longmire in London. "The cufflinks are so flashy that Elton John could wear them, but like the lining of my tuxedo, I like to keep the devil inside."

SIGNET RING

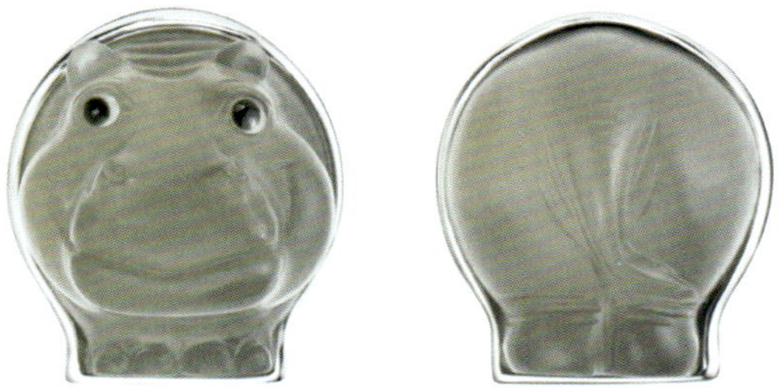

MOONSTONE HIPPO CUFFLINKS

TANZANITE CUFFLINKS

"People will want to be happy again, they will party and enjoy life, and so I assume there will be fireworks of colour."

Constantin Wild

RUBELLITE CUFFLINKS

SHARING THE LOVE

Betonzimmer
Strong Room

SHARING THE LOVE – FAMOUS WEARERS AND COLLECTORS OF GEMS

by Nina Hald

A serious interest in gems is almost the same as writing your name in history. Gems are as close to immortality as we can get, and your legacy will record your name for posterity.

History is filled with royalty, great personalities and famous people who adored gems with the most passionate interest of true collectors. For royalty of many cultures, possession of amassed gems was also a way to legitimise dynastic claims through generations. Some of the best known examples include **Pharaoh Cleopatra** (69-30 BC), who had her portrait engraved on emeralds; **Augustus the Strong** (1670-1733), who amassed one of the most impressive collections of gem-set gentlemen's jewellery; **Tsarina Catherine the Great** (1729-1796) of Russia, who benefited from her adopted country's rich deposits of various gems and the collections of her father-in-law, **Peter the Great** (1672-1725), who had accumulated many mineral treasures in rarity cabinets which she added to; **Madame de Pompadour** (1721-1764), official chief mistress of the French king Louis XV (1710-1774), who dictated fashions as an arbiter of taste; German author and scientist **Johann Wolfgang von Goethe** (1749-1832), who collected minerals, also thanks to his intense appreciation of colours; the last **Chinese empress Tzu-Hsi** (1835-1908), who fancied pink tourmalines to such an extent that she supposedly bought several production vintages of them and was laid to rest with her head on a carved pink tourmaline pillow; and finally, the American businessman and philanthropist **John Pierpont Morgan** (1837-1913), who donated large gem collections to some of the most important museums in the United States.

• • •

MODERN-DAY COLLECTORS

History is one thing – the time we live in is another. What do modern-day collectors ask for when they build their collections? According to Constantin Wild, they want honest advice and guidance on how to set up collections. "Some prefer to assemble a collection of gems meant to be set in jewellery – some gather assortments of museum pieces," he explains.

Collectors display many different approaches to their passion for gems. Some collect by colour (for example every blue stone there is), some by gemstone variety (for example all the varieties of tourmaline), some by country of origin. Some want a specimen of each available species, some want to acquire the best of each variety, some want them set in jewellery and finally there are those who collect all the members of a certain group, like the beryl group, in order to have every shade of beryl represented.

"A collection with a certain theme will always be worth more than the sum of the individual gems, if the group is put together in an intelligent way. Usually, the collectors educate themselves, and they remain constantly on the lookout for sources of information. Some follow the auction houses' advice – however, these will usually focus on the four classic gemstones – diamonds, rubies, emeralds and sapphires – and colour variations thereof," says Constantin Wild.

Also thanks to promotion by auction houses, certain fancy-coloured diamond 'individuals' fetch astronomical prices, but if collectors are willing to look into the whole universe of gems, they discover more interesting, varied and also very beautiful gems that can be obtained at a fraction of the cost of the gems being promoted.

Some collectors keep their gems on display, some in safes. Others carry their treasures around in boxes with them, for example at the Tucson Gem Show, in order to share their love and appreciation of them with like-minded individuals. When one is fortunate enough to be offered a peek, looking inside such a box is like looking at sparkling candy. "Collectors love to show off their gems to me, because they know that I understand and appreciate the true beauty of the gemstones and the work that has gone into amassing them. That is also why I prefer to sell my best gems only to individuals – I know they truly understand and appreciate what they are buying. I have helped serious collectors with advice on how to build their collections in concept and shown them how to present them most advantageously with displays and lighting," Constantin Wild explains.

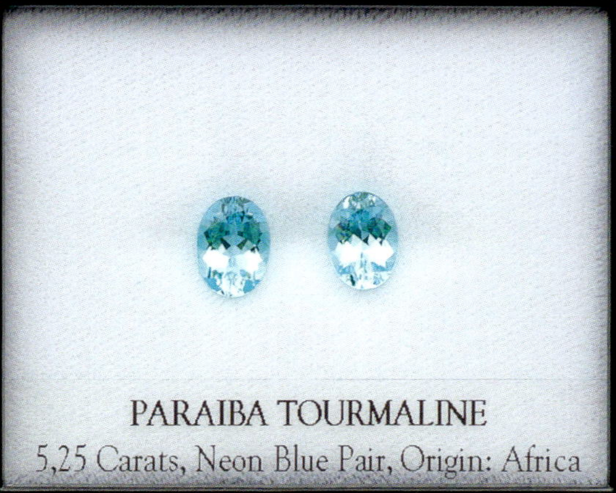
PARAIBA TOURMALINE
5,25 Carats, Neon Blue Pair, Origin: Africa

CANARY TOURMALINE
9,41 Carats, Yellow Colour, Origin: Africa

MANDARIN GARNET
9,08 Carats,Original Mine,Origin:Namibia

PERIDOT
15,80 Carats, Cushion Cut,Origin: Myanmar

MANDARIN GARNET
20,29 Carats, Matched Pair, Origin: Africa

PERIDOT
12,89 Carats, Oval Pair, Origin: Pakistan

SPARKLING INVESTMENTS

In Constantin Wild's experience, after some forty years in the gemstone business, quality gems can be a sparkling investment, but he is also quick to add: "It would not be right to promise an increase in value. However, I have seen all gem types increase in value, and much faster than inflation rates during these past four decades. This is especially true for quality gems – and by quality gems, I mean extremely beautiful, rare and unusual specimens. I am not thinking about the mass-market commercial gemstones, such as small pieces of quartz or blue topazes. I am thinking of gems that are used as centrepieces of important jewellery. I have sold 1-carat Russian demantoid garnets and paraíba tourmalines over the years, and they have increased tremendously in value. When the first paraíba tourmalines from Brazil came on to the market in the early 1990s, they cost about five times as much as a green tourmaline, which at the time was something like 500-600 dollars per carat. Today, the same paraíba tourmalines would fetch more than 100,000 dollars per carat – that is, if you were able to find one. Because they are that rare and that much sought after," Constantin Wild emphasises.

Constantin Wild's approach to luxury is always to examine how much something differs from the ordinary. "When it comes to value, I think in terms of how much of my income I have to pay in order to own a certain gem. When I was robbed of part of my collection of gems in Philadelphia some thirty years ago, I was of course reimbursed by my insurance company. But shortly after, I had to discover the hard way that it was not a real reimbursement, because it was and still is impossible to just go to a market and buy gems of that quality. The collection had been the result of years of collecting," he reminisces.

When a collector wants to buy a certain gem, Constantin Wild will try his best to put a selection together for the collector, over the course of perhaps a month or two, for the person to choose from. But usually there are not that many quality gems on the market to pick from at any given time. Coloured stones are much harder to come by than diamonds. "To find a coloured gem worth a million euros, you perhaps have to search the world to find one or two that fit the given criteria. In comparison, it is much easier to find diamonds that match certain desired specifications. Diamonds have always been highly publicised and promoted, which is also why they fetch much higher prices. But coloured gemstones are the real rarities. So my heart is with the coloured gems, as they display the truest value in beautiful saturated colours. The diamonds are only the roses in the world of flowers, but the flower world has so much more to offer," says Constantin Wild.

Of course, Constantin Wild is a collector himself, and this is both seen and felt in the way that he does not consider gems to be 'just' merchandise. Rather, they make up his treasure trove. The very best of these make their way into his More Than Just Rare collection. Over the years, this collection has entailed about 300 pieces – all highlights of his career, every gem special in some way – because of its colour, size, cut or history. "In the More Than Just Rare collection are also the gems that, in addition to being rare and beautiful, exceed a weight of 100 carats. A dozen or so weighed more than 100 carats, and so they were named the 100-carat club. I have, for example, owned peridots, red fire opals and even a mandarin garnet, each of which weighed over 100 carats. I sold a 107-carat blue moonstone from Tanzania to a private individual, who later donated it to the Smithsonian Institution. At first I was sad but then especially proud to see an imperial topaz that I had acquired myself go to an important American collection. Actually, I loathe parting with my gems. When I see them donated to some of the largest museums in the world with the most prestigious collections, it can make it a little bit easier, but basically, I wish I could keep all my treasures with me," Constantin Wild says.

A genuine collector is not driven by investment interests, or at least not solely. He or she is looking for gemstones and objects that will be of interest for decades and even centuries to come. More often than not, creating a serious collection is difficult, laborious and highly demanding. There is never any predicting when (or sometimes even where) a particular gemstone or object may appear on the market, and the window for acquiring it may be brief. So, when you do come across that perfect gem, do not let it get away.

"GEMSTONES PLAY WITH YOU. WHEN YOU HOLD THEM IN YOUR HANDS, THEY DANCE AND PERFORM - THEY ARE THE STARS IN YOUR LIFE."

Constantin Wild

SEVEN STRONG SETS

Augustus the Strong, Elector of Saxony and an avid patron of the arts and architecture, had perhaps the largest, most varied gem-set collection of gentlemen's jewellery ever seen in the history of mankind – in size rivalled only by those of the Indian maharajahs, though they gravitated more towards diamonds, pearls, emeralds and rubies. One of Augustus's sets was comprised of pieces set with carnelians, one set featured tortoiseshell, one set was made with sapphires, one with emeralds, one with rose-cut diamonds, one with brilliant-cut diamonds and finally one with rubies (although some have turned out to be foiled spinels on closer inspection). The term 'jewellery', in the case of Augustus the Strong, is to be taken in the broadest sense. Each set contained waistcoat buttons, cufflinks and a variety of buckles (for shoes, belts and sleeves); most of the sets included walking sticks with handholds set with gems and ceremonial swords – one set even included a hammer! – military medals, watch chains and tabatières set with gems, all of course strictly coordinated in colours and matching qualities across each individual set. The sky was literally the limit for this dandy, and the gems were either cut cabochon, polished into beads, some engraved, or a great many faceted. These large parures were not intended only for representation – they were also conceived as strategic monetary reserves which could easily be transported through Europe and pawned if needed. In 1719, the gems in the sets were estimated to be worth 3.5 million thalers (one thaler being a silver coin of roughly one ounce in weight, about 25-30 grams). Unfortunately, in November 2019, some of the sets were stolen from the Green Vault (Grünes Gewölbe) in Dresden, where they had been on display for centuries – thankfully, some of the other sets were on display in New York at the time, which kept them safe.

• • •

POMPADOUR'S PIECES

Madame de Pompadour was the mistress of King Louis XV; she was one of the first female fashionistas and trendsetters. At the time, gems were perceived as a way to immortality, because with their eternal beauty they seemed to defy death, and the king lavished them on her. From the inventory of her possessions made at her death, she owned jewellery set with carnelians, sardonyxes, agates, octagonally cut topazes, sapphires and purple sapphires, peridots, emeralds, rubies, diamonds, malachites, opals, moonstones, cabochon-cut garnets and cat's eyes, several important parures with diamonds in many nuances, 40 rings and many cameos. According to tales from court, Pompadour was fond of exceptional colour combinations; an amethyst surrounded by yellow diamonds, a pink diamond surrounded by pale blue diamonds, and blue and green diamonds surrounded by colourless ones. She is said to have owned seven large jewellery sets, mainly set with fancy coloured and colourless diamonds, rubies, sapphires and emeralds, with which she was able to dictate fashions. Many pieces were made to be attached to the large dresses of the times (buttons for sleeves and other ornaments), while she had hats decorated with gem-encrusted garlands and gem-set rosettes. It has been said that Madame de Pompadour herself received training in gem engraving from Jacques Guary (1711- believed to have died in 1793), and that she even attained 'some proficiency' in the field.

• • •

SHOPPING FOR THE RAINBOW

Shelly Sergent, curator of the privately owned coloured gemstone and jewellery Somewhere in the Rainbow Collection with the catchphrase Because a Vault is No Place for Fine Gems and Jewelry, has had the good fortune to view the best gems on the market during the past decades – and acquire many of them for the collection. The latter is used for the expressive purposes of education, exhibition and enjoyment for gemmological organisations of many types, museums, galleries and retail jewellery stores, the ambition being to share the experiences of museum-quality gems and one-of-a-kind handmade jewellery pieces with many organisations, groups and individuals for a better understanding of the rarity of the gems.

The collection was founded in April 2008 with the purchase of a 10-carat alexandrite, a 4.65-carat alexandrite and a 15-carat tanzanite. Today, the collection is home to more than 1,500 gems and jewellery creations, of which 40 per cent are set in jewellery and 60 per cent still loose, awaiting a fabulous designed home. The collection features tourmalines, quartzes, spinels, corundums. garnets, spodumenes, feldspars, zircons, zoisites and many rare gems within their own families, such as olivine and apatite. When asked which gem is first in her mind, Shelly Sergent doesn't hesitate for a second: "The gem known as 'The Eternal Virgin' is an exceptionally large tenebrescent zircon, which turns from red to black and then back to red in two weeks of complete darkness The red is a fire glow red when first seen coming out of the dark. It takes about 7-8 minutes to completely fade to black, and it is a fascinating gemstone."

With gems, Shelly Sergent doesn't 'play favourites'. "Each family of gemstones excites me for different reasons. However, some of my personal favourites are tourmaline and spinel. I love the never-ending colour choices and combinations of tourmaline, and the twinkle of brightness of spinel." When she sees a special gem for the very first time, Shelly Sergent usually has the feeling that it speaks to her. "The colour, the cutting style, the cool inclusions, the story it comes with or maybe who it came from – when a stone moves me, I know it's the right one for Somewhere In The Rainbow. Needless, to say, I'm moved often," she says, laughing. "Sometimes, a gem may be average in the big world of gems, but maybe the gem cutter was a 14-year-old who has taken the road to discovering his own passion. That stone will speak to me because of his story and then I 'get it'. Every single acquisition for me is an exciting journey with a story to be told, and they are all so special that I feel so blessed." As Shelly Sergent is a huge fan of fine German cutting and the legacy behind it, Somewhere In The Rainbow boasts many beloved and exceptional gems from German cutting houses.

"Even if someone has a large amount of money to spend, it can take years, maybe even decades, to find and collect outstanding pieces. This is an absolute truth. Somewhere In The Rainbow is home to some very rare and extremely fine examples of gemstones that, no matter how much money is offered, we could not replace. Mother Nature only gives us so many of these pockets of certain gemstones, and once we have depleted the deposits, there are no more to be found. No more will be produced for many geological reasons, so we must cherish these fine examples and educate the public as to the rarity of these natural occurrences. No matter how much money one has, some gems just cannot be replaced. It's one of the most intriguing and important stories of the gemstone world."

CONCENTRATE ON COLLECTING

Sired by the industry's famous Gübelin dynasty, famed for its jewellers and lab grading, Thomas Gübelin, who is a friend of Constantin Wild and a gem collector himself, has the following to say about collecting gems.

"Start out by collecting whatever gem you please. Select whatever appeals to you. After a while you find that you have collected a lot of gems of different species, colours and sizes, and that they amount to a real mess. That is when you start concentrating on finding your own collecting strategy. You have got your feet wet, and it has given you an inkling of what appeals to you; is it actually the species, the variety, colour, cut, size, country of origin, phenomena – or perhaps something else entirely? Only you can decide. I myself like chrysoberyl cat's eyes, garnets and spinel, because of their varieties in colours and the phenomena."

"When I began in the business, I didn't know much about gemstones. As a jeweller I knew what I needed for the pieces that I was working on, and I knew the golden rules of thumb – 'never put emeralds or opals in ultrasonic cleaning', for example. Then I studied at the GIA, and that is when I started to love gems. The chrysoberyl cat's-eye's movement over a cabochon-cut stone is eternally fascinating for me. I am forever looking for cat's eyes, also in spinels, garnets and quartz. Stars as well. I go crazy for them. When I try to explain my fascination with the phenomenon known as the cat's-eye effect, I always recommend that people place the cat's eye on a grey board. Perpendicular to the eye stone, light two candles, one on each side. Then start turning the board 90 degrees, and see for yourself how the eye opens and closes."

"I like to think that sooner or later, a gem that I am meant to own will usually find me. When I go to the industry fairs, I am not searching for a particular gemstone. I just walk around – window shopping if you will. But suddenly, something looks at me! It catches my eye, and my attention. I usually fall in love with it – and I wasn't even searching for that particular gem. If by chance somebody else acquires the stone that I wanted, then I am happy for the owner. I am never disappointed. Others should have the opportunity to acquire gemstones as well, and I am always happy to give them that opportunity. However, I had to acquire a magnificent 3.46-carat canary greenish-yellow tourmaline with an extraordinary brilliance from Constantin Wild because of its perfect oval cut. I simply couldn't pass that up, but had to own it."

"I believe it takes as much patience and time to build a fine collection of coloured gems as it does to build a collection of diamonds. It you want a rainbow collection, it can take an amazingly long time to find and evaluate both the diamonds and the coloured gems. Both could easily take a lifetime to put together, but it is also a very rewarding task. A collector never really finishes collecting. There is always something new to discover within your particular passion. Talking with others about what Mother Earth has to offer, showing them what you have collected and being fascinated alongside them gives me endless pleasure," Thomas Gübelin concludes.

IMPERIAL TOPAZ SET

175 carat

QUALITY IS QUALITY

To collector Amanda Rubin, who has purchased gems and jewellery for the past twenty years, there is no doubt: colourless diamonds rarely interest her, but coloured gemstones such as paraíba tourmalines and other coloured tourmalines, padparadscha sapphires, spinels – especially the hot pink ones, but also silver spinels – imperial topaz, opals, emeralds and brightly orange spessartines are the gem loves of her life. "I recently purchased a stunning 10-carat malaia garnet that I fell in love with," says Amanda Rubin.

AMANDA JOY PARAIBA

Her approach to collecting is a straightforward one. "I go with my heart and what pleases my eyes. I work from a wish list of stones that I want to add to my collection. Currently I am looking for a neon green diamond – I let one get away, but I have never forgotten it – a really good padparadscha, a neon blue spinel and a red imperial topaz. However, I am always open to what I may find that is particularly exquisite, and when I come across something like that, I try to purchase it if I can," says Amanda Rubin.

Looking for good value is not always easy, because – as Amanda Rubin has been forced to realise – the best and most exciting stones continue to go up in price. "When I visit Constantin, we talk about stones he had that I wasn't able to purchase, and he remembers them too. There was a paraíba that was so beautiful and such a radiant blue that the entire room seemed filled with its light. It was many years ago, but at the time, I couldn't afford it," she says. Amanda Rubin will usually buy gems with certifications, always from reputable dealers, and more often than not, she has a gemmologist look over the stones first, just to be sure.

All the gems bought by Amanda Rubin will be set in jewellery sooner or later. "Some of my gems are just waiting for the right piece to be set in. I especially like rings, because I can appreciate them while wearing them and that gives me great pleasure. One friend once chastised me about spending money on gems, but recently she actually apologised to me, because even she could see that the prices kept going up, or at least the gems would hold their value. Quality is quality. For many years I made do with ogling Constantin Wild's beautiful acquisitions at the Tucson fair. He has always had a stellar reputation, and when you walk into the German section of the GJX show at the Tucson Gem Show, you know you are meeting some of the ultimate dealers with some amazing stones," says Amanda Rubin.

"Imperial topazes appeal to me because some look like they've got a fire inside them. When I had a big birthday coming up, I decided to purchase something extraordinary for myself. My stone is orange and red and yellow, and it is just exquisite; I can get lost in its beauty, and I had the designer Cyril Despointes set it in a ring for me," Amanda Rubin concludes.

AMANDA JOY TOPAZ

THE DRIVE TO ACQUIRE GEMS

Gem collector Mehboob Hamza has a very particular approach to colour in that he likes to link natural gems with music, birds, fish and flowers. With a background as a Wall Street analyst and entrepreneur since the age of 24, and several successful computer start-ups behind him in the United Arab Emirates since 1983, Mehboob Hamza feels driven to collect the entire range of gems.

"I look for the unusual; a blue tourmaline that looks like a cornflower-blue sapphire – that appeals to me because of its rarity. When I look at gems' colours, they must speak to me, instantly, like music. Colours in gems can display dimensions such as colour depth, saturation and dispersion which no camera can capture, only the human eye," says Mehboob Hamza. He considers himself a romantic when it comes to gem collecting. The romanticism turns into impulse, because just as with beautiful poetry, music and painting, the whole brain becomes involved and excited when looking at gems and acquiring them. "Collecting gems is a year-round activity for me. Even on a daily basis, I get pictures from merchants and phone calls about particular gems that are up for sale. I work closely with 12-15 dealers around the world. When they come across a particular rare rough, they know to call me. It is a 'drive' in me to acquire gems; I never wait for a special occasion."

Mehboob Hamza's grandfather was one of the top gemstone dealers in Sri Lanka in the 1960s. Suppliers of roughs would come to him in the morning, have tea and breakfast with him, and then, at 7.00 or 7.30, he would look at the gems because the light was right. "I have been holding gems in my hands since the age of 5, and I was the only one in my family interested in gems besides my grandfather. Thanks to these childhood experiences, I can see certain colours that even gem dealers can't see. I have a colour reference that they don't possess. That is also what makes it easy for me to compare gemstones to the colours of animals. When you see the Indian peacock's neck in the wild, in different shades of light, it is fascinating to behold. It gives you dimensionalities of the colours that you don't see otherwise," says Mehboob Hamza.

Mehboob Hamza collects gemstones because they give him a feeling of nirvana. "Whenever I have spare time, I take my gems out of my safe and I look at them for several hours. The effect of gems is rather like that of music; if you keep listening to the same piece, you start listening to the intricate harmonies and contrapuntal melodies that may not be on the forefront. Listening to the same piece of music constantly gives new pleasures. The same happens with gems. The more I look at them, the more I appreciate them. Because of different lighting during the day and night, gems will display different dimensionalities of their colours."

Among the most important gems that Mehboob Hamza has acquired from Constantin Wild, he recalls an incomparable cuprian tourmaline that he had set in a ring for his wife. "Another gem from Constantin that I remember clearly was a fairly large-sized Mexican opal with an 'egg' inside, which is sort of equivalent to a horsetail inclusion in a demantoid. Some time ago, I bought a red topaz from Constantin. The price point was incredulous, but on closer examination on its rarity and beauty it was a no brainer, so I had to have it. When I see something so unusual, it gives me an adrenaline kick. Those are the experiences that remain in my memory. A 52-carat ruby-red loupe-clean spinel; that was irreplaceable. A multicoloured 8-carat sapphire from Madagascar – the colour dispersion was all over the stone in a collage of blue, green, orange and yellow. I made an offer, but I missed that stone, and that still haunts me," he says.

"When I speak with Constantin, it is clear that the gems are never just merchandise to him – they are the pleasure of life. Constantin knows my taste and what I look for, which in turn helps me a lot. I don't have to see 100 gems; I just need to see maybe 10 or 12. There is a trust between us," Mehboob Hamza concludes.

BLUE TOURMALINE

17 carat

SPINEL

5 carat

YELLOW & BLUE TOURMALINE

42 carat

AMETHYST

38 carat

COLOURS IN CHINA

Francis Cui, managing director of HSTC Private Jeweller, was appointed ambassador of the ICA (International Colored Gemstone Association) to China in 2012. As an experienced gemstone and jewellery expert with Graduate Gemologist and Accredited Jewelry Professional degrees from the GIA, he contributes to the luxury magazine Robb Report amongst others, and he is often invited to give lectures and seminars by private banks, clubs and other organisations. As part of his mission to bring the rarest and most beautiful coloured gems to China, Francis Cui is working on the creation of a gemstone and jewellery museum which will highlight the multiple facets of the mysterious world of gemstones.

Francis Cui's own favourite coloured gems are ruby, sapphire, emerald, and red beryl, and since childhood, he has collected several thousand carefully selected gemstones and gem-set jewellery pieces. He builds his personal collection by acquiring gems from every imaginable source, including mines, cutting factories, dealers, auction houses and private individuals. "I base my decisions to buy on the beauty of a stone – its colour, crystal formation and cut. I buy the gems that attract my eye at first glance and touch my heart after carefully checking. I have bought many from Constantin Wild, and I should mention one amethyst in particular. It was of a good size, with a very saturated red-purple colour, which distinguished it from most of the amethysts to be had on today's market. Constantin told me it was from the company's very old gemstone collection. It might have been acquired by his grandfather or even great-grandfather a long time ago. I created a beautiful ring with this gorgeous stone in my Paris workshop. It was displayed in my boutique shop in Beijing, but soon after, it was acquired by a nice lady who had travelled all the way from Paris. "What a coincidence!" says Francis Cui.

When asked to characterise the gems of Constantin Wild, Francis Cui says: "Constantin Wild's gems are always saturated with bright colours, and they display the best cutting skills and a magic attraction. I believe the beauty of his gems is created by his good eye, nice taste and passionate heart."

• • •

THE SCOTT COLLECTION

In the 1980s, Apple Computer's first CEO, Michael Scott, became intrigued by gems when he discovered that a stone, passed off as a 5-carat tsavorite for a ring he was planning on purchasing, was actually a man-made stone – a so-called YAG. That YAG was only worth a fraction of the amount a similar-sized natural tsavorite would have cost. The experience made Michael Scott devote himself to learning more about gems, even becoming an expert through the collecting process. Since then, he has spent decades putting together one of the world's largest and finest collections of gemstones. As he sums it up in the introduction to the catalogue accompanying the exhibition 'Light & Stone', published in 2002 by the Bowers Museum in Santa Ana, California: "My goal has been to collect quality and beauty ... and for the collection to tell the emotional story of the impact of colours."

When highlights of the Scott Collection were shown at the Bowers Museum, the museum's president Peter C. Keller termed them "world-class". The sentiment was shared by CEO William Thorsell, of the Royal Ontario Museum, in 2008-2009, who dubbed the selection "truly peerless" when it went on display there. The Scott Collection has been called the most important private gem collection in the United States, with few rivals in the world outside the royal families. The collection displays the entire colour range, in sizes and qualities rarely seen. In addition to rough and faceted gems, the Scott Collection also includes many magnificent examples of gem sculptures. When it was on display at the Shanghai Museum for six months in 2005, it is said that almost a million visitors saw it.

"Collecting the world's largest and best examples of gem species has been my passion for almost 25 years," said Michael Scott in 2008, when the exhibition opened at the Royal Ontario Museum. Michael Scott has even been given the highest honour in gemstone circles: the mineral scottyite is named after him.

• • •

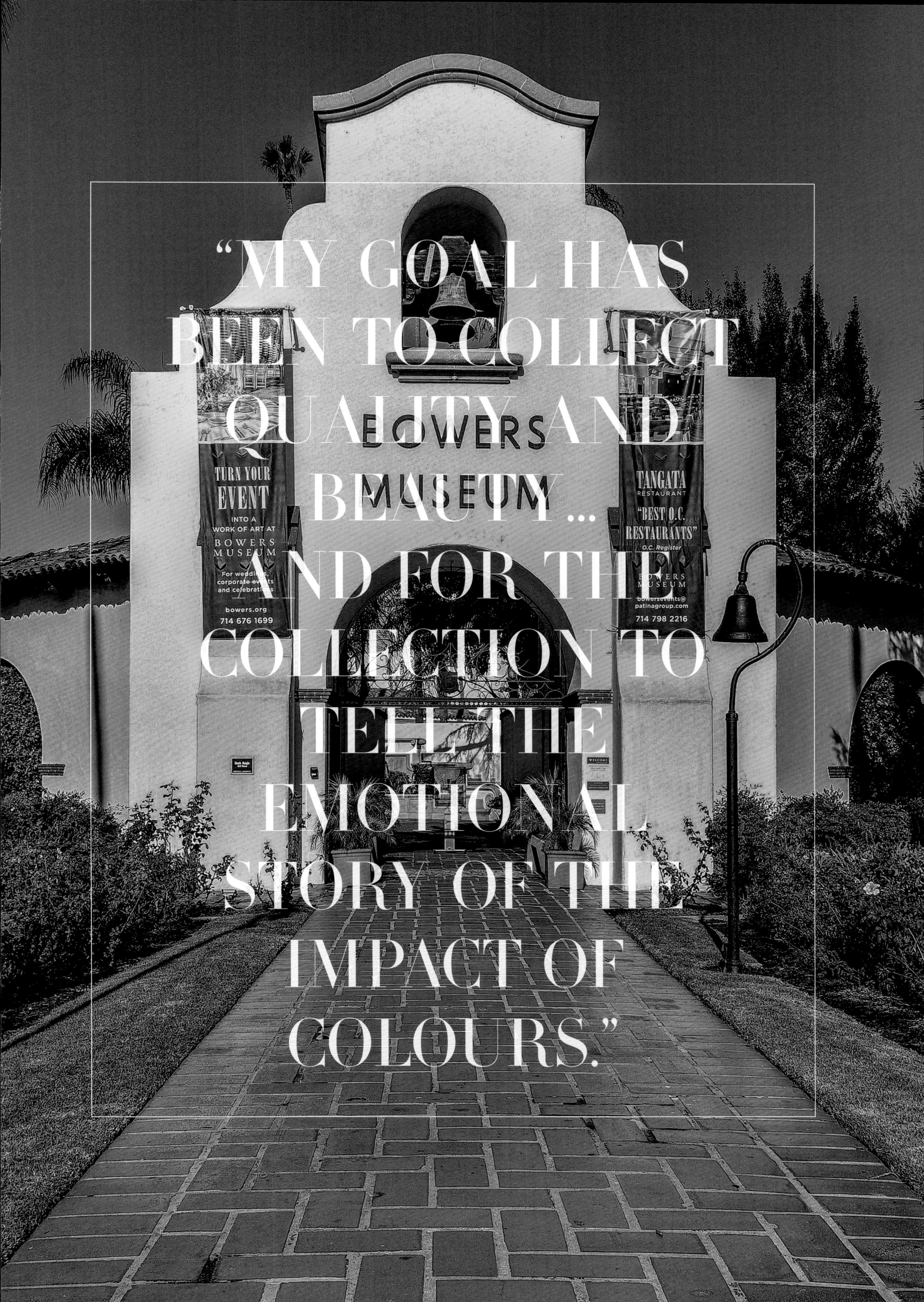

"MY GOAL HAS BEEN TO COLLECT QUALITY AND BEAUTY ... AND FOR THE COLLECTION TO TELL THE EMOTIONAL STORY OF THE IMPACT OF COLOURS."

Photo: Fred Leighton

Gemstones as luxury

by Nina Hald

The definition of what luxury is has changed over time, but gems have always been regarded as pure luxury since the early days of mankind.

Travelling in style and staying at splendid seven-star hotels. Owning superlative homes and gardens. Driving great (fast!) cars. Eating the best Michelin-star food with vintage wines. Bags, clothes, perfumes and phones; each consumer category has a luxury level. To communicate their exclusivity, the producers of these goods seek affiliation with the gem and jewellery world.

Take a 'platinum' credit card for instance. Sapphire status membership. Or car manufacturers naming their prestigious colours turquoise blue, emerald green or ruby red. Gemstones set the tone for the highest pinnacles of luxury.

Gemstones serve various staging purposes and effects. Rulers have broadcast their majesty by jewellery set with gemstones. Priests have flaunted rings, meant to be kissed in faithful obedience. Even in the pre-Chinese military, rank was communicated with jade. For that reason, gemstones became symbols of sovereignty and power – the most prized possessions in every civilisation.

Exceptional gems set in jewels, for example those in the British crown, still evoke the power and history of generations past.

IMPERIAL TOPAZ

3o carat

TSAVORITE GARNET

8 carat

TANZANITE

21 carat

YELLOW SAPPHIRE

26 carat

SPINEL

11 carat

Luxury is desire and pleasure, related to our dreams and fantasies. Luxury is defined as something extraordinary, limited in availability and aimed at the privileged few.

As societies have developed, gems have always remained luxury objects due to their rarity. Gems were picked up from the ground because of their colours and crystal lustre, distinguishing them from other rocks and minerals. At a time when bright, vibrant pigments were difficult to obtain and preserve, gems enthralled and endured. Gems have always been admired and collected. They have a timeless appeal and value and can be passed on down to the next generation.

Excellent-quality coloured gems are extremely rare. In his 40 years in the company, Constantin Wild has only seen prices rise, much faster than inflation rates.

"If you want to buy a coloured gem, do it now! There is no time like the present. However, trying to buy the loose gems yourself on the world market is risky, especially if you are not familiar with quality. Each variety is judged by different criteria – some on size, some on brilliancy or colour, some for being as flawless as possible, others for having the right inclusions. You need a specialist with you to be able to evaluate a good buy. To be on the safe side, purchase a quality gem set in a piece of jewellery from a leading brand. A prestigious brand will give you certainty for the quality of the stone," Constantin Wild advises.

■

THE GEM HOUSE
W.C.Wild & Co.

THE GEM HOUSE – Tour de Wild

by Nina Hald

A privileged tour of Constantin Wild's historic headquarters provides exclusive and elusive insights into the world of coloured gemstones.

The powdery pastels of lavender sapphires and bluish moonstones. The powerful presence of canary and neon blue paraíba tourmalines. The lushness of green peridots and red spinels. The gems of Constantin Wild inspire moods of tenderness, playfulness and festivity. Special orders, romantic escapades and the pure joy of gemstone magic are all housed here in the vaults of Constantin Wild. His veritable treasure trove of the 21st century is enough to leave anyone breathless. And even the most devout diamond-lovers will find themselves making bids for his riches.

For ten generations, the Wild family in Idar-Oberstein has been working with either gem cutting, jewellery making or gemstone trading. Without interruption. Even today, the family premises stand on the very same idyllic plot of land: a calm cherry oak garden in the centre of the quaint overgrown village. This is where the old cutting mill, the 'Kirschgartenschleife', stood. Next to it was the house of Constantin Wild's great-grandfather's family, turned into the current building in 1911-1912. At that time, the old cutting mill was pulled down, and the old family house converted into a fine, white, palatial villa. At the back, the farm's old stables for stallions were converted into an office building that was modern at the time. Constantin Wild's father, Fritz Wild, added to the building in the 1950s, but today the estate remains as Constantin Wild's grandfather built it in 1911-1912.

Over the past forty years, Constantin Wild has undertaken a renovation of the entire building. During that process he also acquired ownership of all parts of the building previously held by various members of the Wild family.

Blue moonstone cabochons

next page:
top left: Main street, Idar.
bottom left: Architectural drawing of office building at Hauptstrasse 103, W. Constantin Wild & Co., 1914, by H.Weszkalnys.
top right: Architectural drawing of private house at Hauptstrasse 103, W. Constantin Wild & Co., 1911, by H. Weszkalnys.
bottom right: Office building of Constantin Wild GmbH & Co.KG, c.1940.

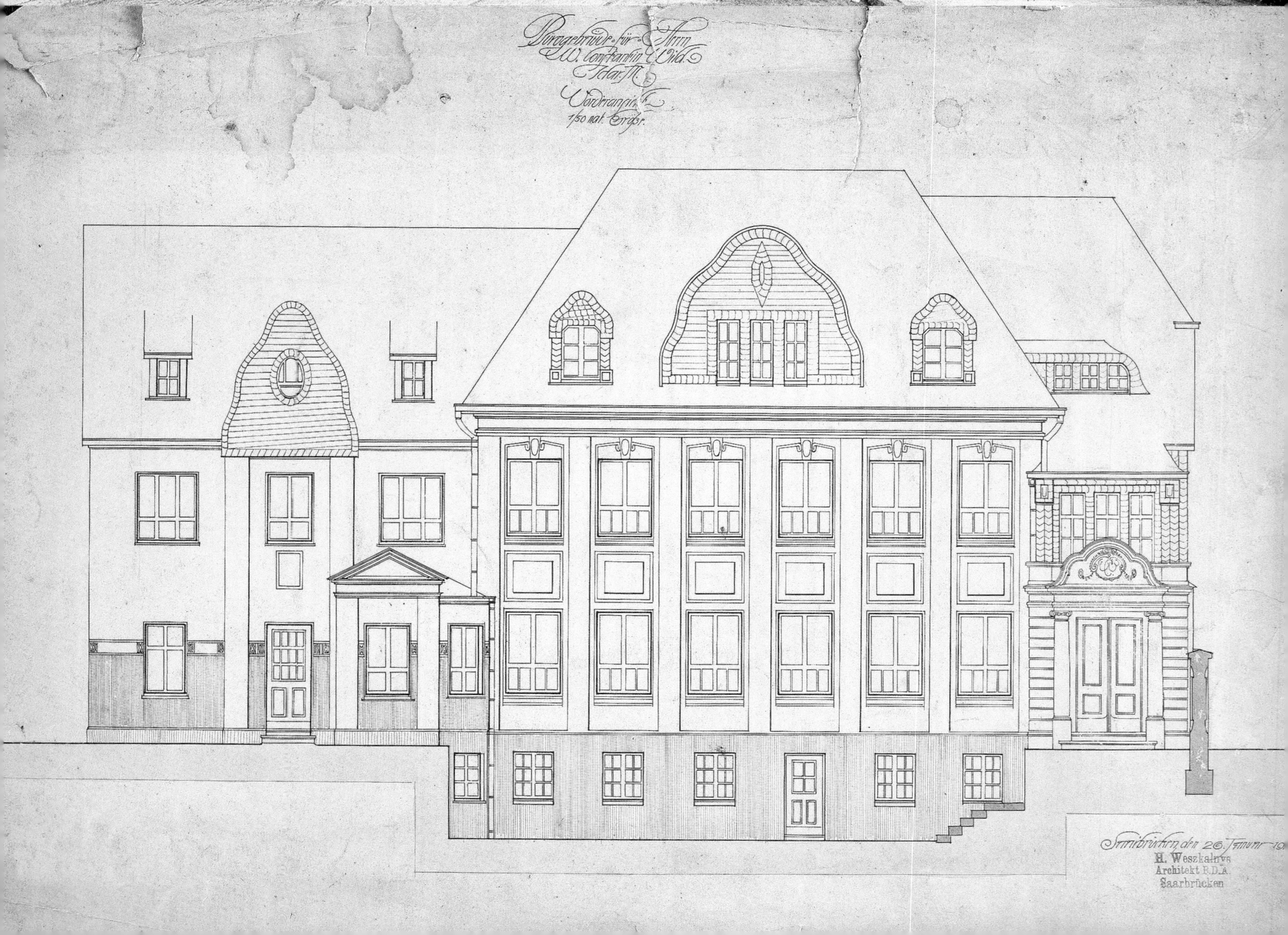

1/50 nat. Größe
Saarbrücken den 26. Januar 19
H. Weszkalnys
Architekt B.D.A.
Saarbrücken

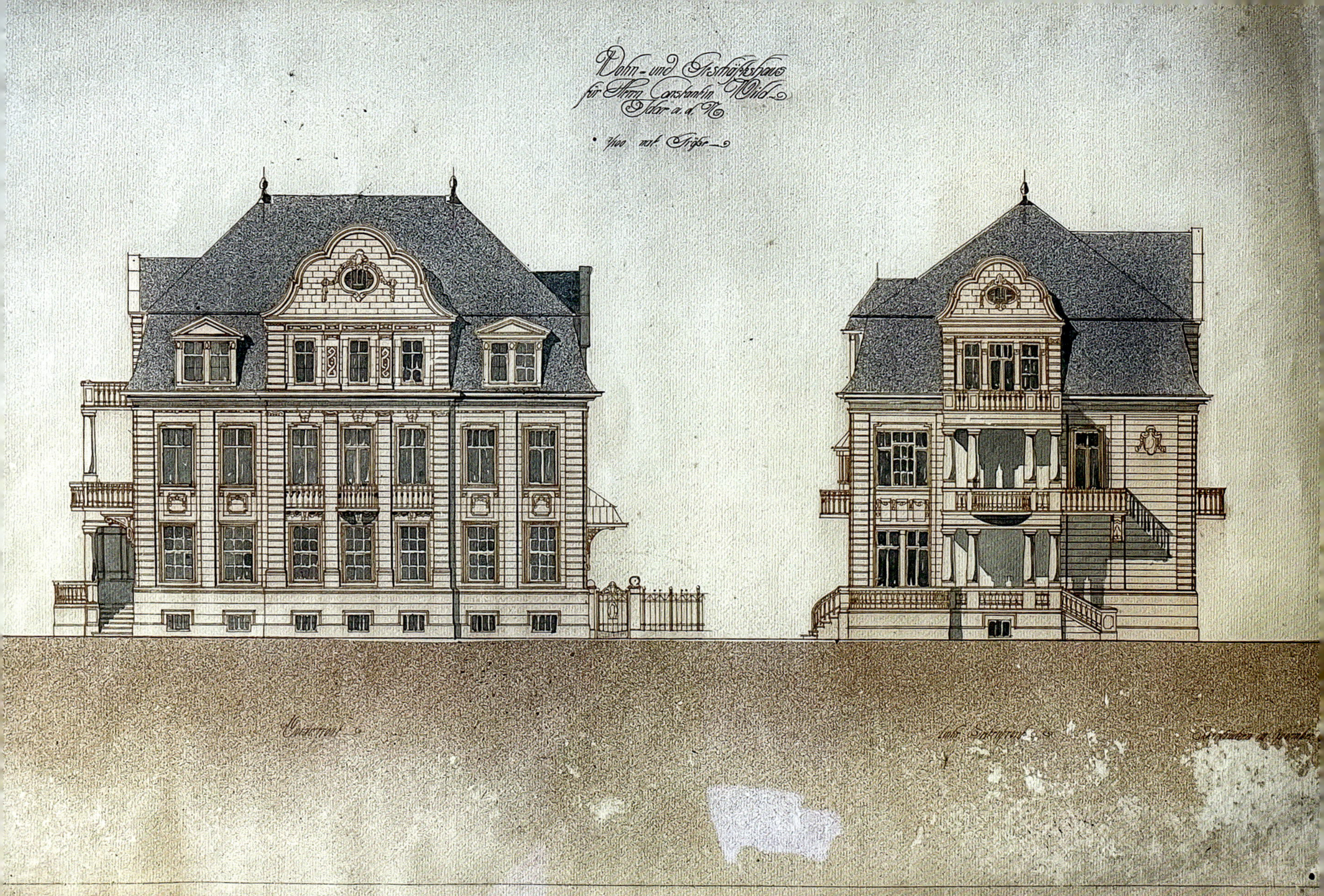
Wohn- und Geschäftshaus
für Herrn Constantin Wild
Idar a. d. N.
1/100 nat. Größe

Rosenquarz
Saphir
Chrysopras
Peridot
Granat
Kunzit
Hessonit
Türkis
Labradorit
Rhodonit
Amazonit
Flussperlen
Perlen
Perlen
Feueropal
Smaragd
Amethyst
Rosenquarz
Aquamarin
Edeltopas
Bergkristall
Beryll
Diamant
Chrysoberyll
Turmalin
Achat
Amethyst
Lapislazuli
Rhodochrosit
Blutstein
Sonnenstein
Blutstein
Bernstein
Achat
Jaspis
Nephrit
Chrysopras
Granat
Opal
Citrin
Korallen
Edeltopas
Demantoid
Aquamarin
Amazonit
Diamant
Malachit
Alexandrit
Aventurin
Smaragd
Rhodonit
Amethyst
Lapislazuli
Grossular
Aventurin
Türkis
Jadeit
Perlen
Türkis
Heliotrop
Perlen
Lapislazuli
Almandin
Rubin
Saphir
Karneol
Moosachat
Diamant
Spinell
Smaragd
Zirkon
Hyazinth
Aventurin
Diamant
Peridot
Malachit
Calcedon
Diamant
Beryll
Turmalin
Aquamarin
Granat
Smaragd
Tigerauge
Rosenquarz
Diamant
Türkis
Granat
Aquamarin
Beryll
Turmalin
Granat
Amethyst
Citrin
Bergkristall
Rosenquarz
Amazonit
Achat
Labradorit
Kunzit
Rubin
Saphir
Chrysoberyll
Spinell
Zirkon
Hyazinth
Mondstein
Almandin
Korallen
Perlen
Opal
Saphir
Diamant
Nephrit
Rohstoffquellen der Fa. W. Constantin Wild & Co. Idar-Oberstein

ALLENBACH
Die „Schleifen" im Idartal
vor 100 Jahren.
LEISEL
1 Kudis Schleife
2 Kässchmierbüchs - Schleife
3 Klinge Schleife
4 Gordners Schleife
5 Petsche Schleife
6 Bollers Schleife
7 Kellers Schleife
8 Beckers Schleife
9 Katzenlocher Schleife
10 Weiher Schleife
11 Obere Preussische Schleife
12 Untere Preussische Schleife
13 Obere Neumanns Schleife
14 Untere Neumanns Schleife
15 Fuchse Blutstein-Schleife
16 Bären Schleife
17 Bauern Schleife
18 Bohnedippe Schleife
19 Hanspeter Schleife
20 Obere Sägemühl Schleife
21 Untere Sägemühl Schleife
22 Leysers Schleife
23 Hohljobs Schleife
24 Bitze Schleife
25 Purpers Schleife
26 Algenrodter Schleife
27 Kirschwiesen Schleife
28 Dreherpeters Schleife
29 Obere Staden Schleife
30 Mittlere Staden Schleife
31 Untere Staden Schleife
32 Schleife im Dorf
33 Schleife unterm Dorf
34 Obere Hettsteiner Schleife
35 Mittlere Hettsteiner Schleife
36 Untere Hettsteiner Schleife
37 Ölmühlen Schleife
38 Hettsteiner Mühlen Schleife
39 Chaussee Schleife
40 Ostindien Schleife
41 Kellwiesweiher Schleife
42 Kotges Schleife
43 Oberstgarten Schleife
44 Unterstgarten Schleife
45 Oberst Wörth Schleife
46 Unterst Wörth Schleife
47 Doktors Schleife
48 Kirschgarten Schleife
49 Mühlen Schleife
50 Dietzen Schleife
51 Kuhnen Schleife
52 Kaulwiesen Schleife
53 Hochsteg Schleife
54 Papiermühlen Schleife
55 Burgmühlen Schleife

Kirschgartenschlei

W.C.Wild & Co.

Located next to the historic gemstone bourse, the 'Deutsches Edelsteinmuseum' (German Gemstone Museum) and the Diamond Apothecary, the Wild villa holds pride of place. The main street of the town – Hauptstrasse – is also known as Germany's 'gemstone mile'. If you want to lay claim to fame in the gemstone business, you need to be represented on this stretch of road. And of course, Constantin Wild and his ancestors have owned the most prestigious square metres on it for more than a century.

'Kirschgarten' cutting mill, c. 1905.

The old water-powered cutting mill may have been demolished in 1911, but large millstones from times past have been preserved. Today, they line the driveway entrance. A small bubbling fountain in front recalls the calming sound of the Idar stream at the back, and a birch tree provides some shade to the front courtyard.

The premises today are composed of two buildings, with gemstone cutting still being done at the back of the villa.

The column-framed historic entrance stands under the sign of 'W. C. Wild & Company' written in golden italics. Through heavy wooden doors, in the small antechamber, the visitor immediately enters the world of gemstone cutting. Inside this windbreak, a mural displays a map of the Idar stream with its 55 historic water-powered cutting mills. The area was considered heavily industrialised a century ago. The 'Kirschgartenschleife', the Wild family's cutting factory, is highlighted as number 48 on the hand-drawn map. On the opposite side, a painted glass mosaic window depicts the old cutting mill, a further reminder of the traditional craft which took place in the villa.

left, clockwise:
World map of W.Constantin Wild & Co. gemstone resources.
Leaded window drawing of the 'Kirschgarten' cutting mill.
Historical lettering on the company house entrance.
List of water-powered cutting mills along the Idar stream, c. 1900.

In the reception area, a huge painted map of the world, prominently situated behind the leather-embossed counter, indicates the locations of major gemstone deposits, some even centuries ago – and all suppliers to C. Wild.

From the reception area, Constantin Wild invites privileged guests on a tour, beginning with the villa's basement. Here, piles of rough gemstones from past generations are kept. In wooden crates and huge boxes, large chunky roughs are seen – for example rose quartz from Madagascar, nephrite jade from Russia, lapis from Afghanistan, rock crystal and agate from Brazil, petrified wood from North America, opals from Australia, haematite from Cumberland in England and jasper from Constantin Wild's own mine.

Along the walls, large potato sacks filled with moss agate, aventurine and amethyst are lined up. On the opposite side of the room – and do mind your head, or you may bang it on the rather low roof – decorative animal carvings, artistic bowls, Easter eggs, ashtrays, chess sets, bead necklaces and trophies are on display. These may no longer be an important part of Constantin Wild's range; nevertheless, they are proof of previous generations' diligent work. Onyx rings, cabochons, intaglio, cameos and more are kept in cabinets along the wall, and even old-school hand-turned lapidary cutting tables and cutting mills can be seen alongside the tools that were used, for example, to separate rock crystal from amethysts.

A lapidary in the cutting workshop.

Making our way back up to the first floor, we enter the cutting workshop. Here, all the new computerised machines are lined up and turning under the supervision of today's workshop head. This, indeed, is where a great deal of the magic happens. Here, dreams and plans for gemstones are carried out – unveiling the soul of the stones. Like ghostly echoes from the walls, age-old sounds of gemstones meeting rotating wheels and discs are heard. These are rooms of quiet concentration and contemplation, intense work and the unveiling of enormous beauty.

right, clockwise:
Computer-aided gem design.
Preformed gems on dops.
Hammering of rough amethyst.
Rough stones in the company basement.

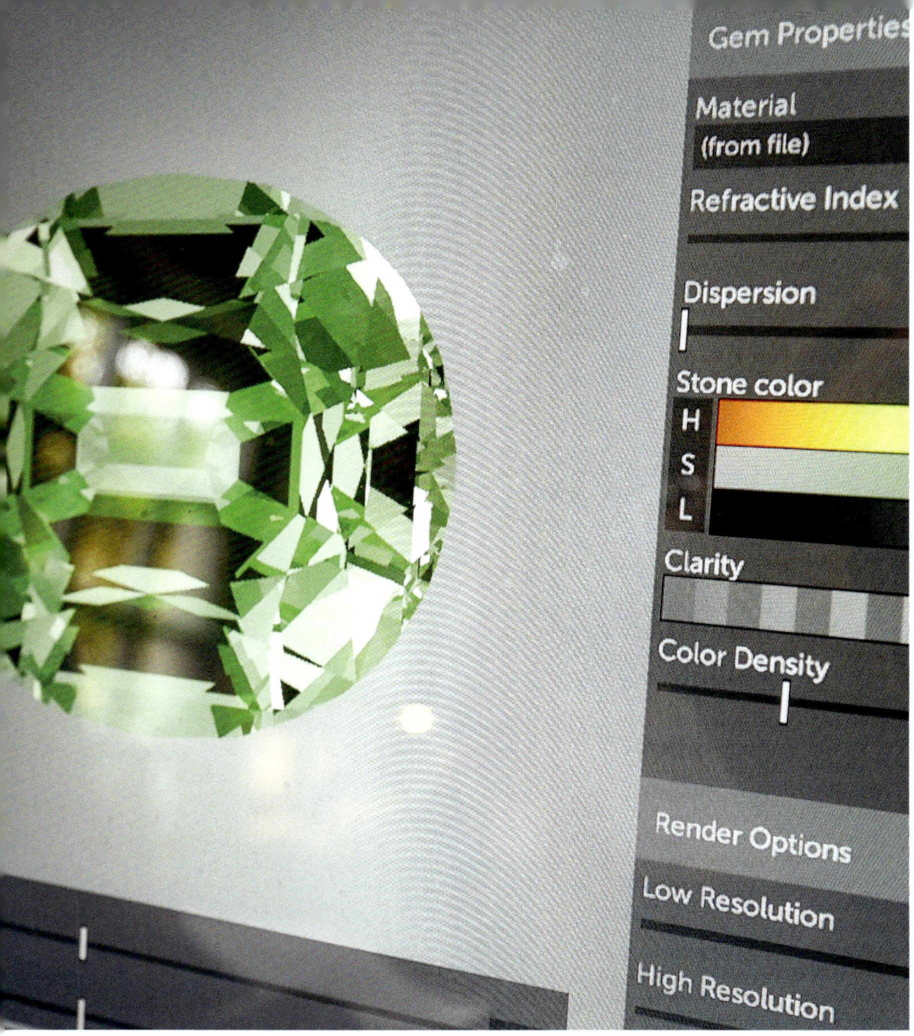
Gem Properties
Material
(from file)
Refractive Index
Dispersion
Stone color
H
S
L
Clarity
Color Density
Render Options
Low Resolution
High Resolution

Schleiferei
Lapidary

Moving back to the main part of the company building, we enter the office suites of Constantin Wild. "This is where the cows were kept during my grandfather's time," the eponymous owner explains. Next to Constantin Wild's desk, the wood-panelled wall is lined with a showcase that houses historic examples of the company's records, drawings and old photographs, rough gemstones and mineral samples from all over the world – some of such outstanding quality that today's finds in no way compare – old passports and travel documents from Wild's ancestors, diplomas of various feats and from various fairs, thank-you notes and congratulatory letters from famous clients, rarities and curiosities from the gemstone world gathered over centuries, antique scales for weighing cut gems, objets d'art and old sample boxes of cutting styles. The dust rests heavily on the items that fill the taffeta-lined trays of the display cabinet.

"I've added a few things to the showcase. I've removed a few things too, but I never organised them out of respect for my ancestors. Perhaps, the day I retire, I will go over everything in there. But then again: I never want to retire," Constantin Wild says, leafing through some of the records.

One of the additions that Constantin Wild has made to the cabinet of curiosities attracted considerable national and even international attention: a sample box of gemstones that arrived at Idar-Oberstein in a spectacular way. Gemstone and jewellery writer Cynthia Unninayar summarised the story for the International Colored Gemstone Association's *InColor* magazine ...

A few years ago, Constantin Wild received a package from Pakistan. "When I opened it, I could hardly believe my eyes. Inside was a small leather-bound booklet containing a mini-collection of gems which had once belonged to my late grandfather, Wilhelm Constantin Wild," explains Wild. The collection – a booklet with 29 rough gemstone specimens – was found off the coast of Pakistan, not far from the port of Karachi, in a bag wrapped in plastic. The fisherman neither asked for nor expected any reward for returning the book, but Constantin Wild gave him a thousand dollars as a token of his appreciation. "My family has been trading and processing gemstones worldwide since the 16th century. So, whilst this was a surprising place for the gemstone collection to be found, it was typical of my grandfather, who learned about the opportunities a global market offered from his own forefathers. He had sample booklets made for selected customers, either as specimens or gifts," Constantin Wild explains, delighted that a piece of his family's history and a specimen of those former marketing materials had found their way back to Idar-Oberstein.

Back at Hauptstrasse 103, amid wooden panels, a heavy oil portrait of W. Constantin Wild gazes at visitors from behind his grandson's chair. Here and there, on the walls, silver-coloured wooden-framed diplomas testify to the company's illustrious past; how ancestors participated in world fairs, gained honours and won competitions. There are congratulatory letters from chancellors of Germany on festive occasions and even scans of ledgers with important sales.

right, clockwise:
Oil portrait of W. Constantin Wild 1938.
Diplome de Médaille D'Or at World Exhibition, Paris 1937.
Gem specimen collection distributed by W. Constantin Wild in the 1950s, returned in 2017.
Malachite sculptured by Nature.

ÉPUBLIQUE FRANÇAISE
SITION INTERNATIONALE
S ET DES TECHNIQ
PARIS 1937
DIPLOME
DE MÉDAILLE
ÉCERNÉ À :_ W. Constantin Wild & Co., Idar-Ob
Allemagne.
CLASSE 55 bis.
LE COMMISSAIRE GÉNÉRAL,
LE MINISTRE DU COMMERCE,

PRECIOUS STONES - GEMSTONES
Garnet
Tourmaline
Constantin Wild
geb. 1866

the value
of colour

Betonzimmer
Strong Room

GEMS. TRUST. WILD.

When we emerge from Constantin Wild's humoristically self-proclaimed 'Wild chamber of power', it is but a few steps to the Strong Room. Here, iron, concrete and steel – supported by surveillance cameras and other high-tech security equipment – house the heart of the company, that which Constantin Wild loves: the gems. In front of a black concrete wall, built to look like a giant bank safe with numbered drawers and compartments, gems are on display – in all imaginable colours and numerous sizes, shapes and forms of loveliness. Inside, the colours of the gems are as breathtaking, saturated and gorgeous as any connoisseur could dream of. After all, this is where fantasy meets reality in miracles made by Nature and perfected by man.

AQUAMARINE

13 carat

The gems are organised in categories established by Constantin Wild. Aquamarines and paraíba tourmalines are grouped together for a pastel impression. Red spinels with the green tsavorites for a contrast of colours. Canary tourmalines with paraíba tourmalines to show the versatility of the tourmaline family. Demantoid garnets next to the spinels because of their high lustre and refraction. Opals next to blue moonstones. Peridots displayed together with amethysts. Soft morganite pink next to the kunzite, and so it goes on. On a good day, millions of euros are on display in the Strong Room, condensed into a few dozen stones – amazing wonders of Nature. This room is truly the heart of the company. The gems rest in the Strong Room, just after leaving the cutting workshop and before they begin their global travels.

The Strong Room at Constantin Wild leaves all gemstone lovers slightly dazed, confronted with unimaginable beauty. But right next to it, a small staircase leads upstairs to ...

left:
Impressions from the Strong Room.

... the Gem Room. A space filled with light. On a rotating showcase we see an amazing amethyst – pure purple power. A single spinel boasts a deep carmine red. In a cabinet, an imperial topaz shimmers. A rough green beryl specimen of unheard-of size is nonchalantly displayed. In each nook, corner and cranny, gemstones bask in the light. Some single spectacular stones. Some matched up in pairs. Some in select suites. Their facets shine and sparkle. In built-in showcases along the walls, the glamour of superlative coloured gems is on display; this is where the *crème de la crème* is kept, where the pride and joy of Constantin Wild is exhibited. To those who for any reason did not get their fill in the Strong Room, the Gem Room is heaven. Deep leather armchairs are placed near the windows for anyone who wants to look more closely and admire the perfectly cut gems.

In an almost merciless quality selection, Constantin Wild deals only with the top three per cent of the world's gemstone production. Of those, only the very best gems of each category make it upstairs to the Gem Room. And fate will have it that some of the impressively large specimens have become permanent fixtures. Constantin Wild cannot let them go. "A buyer would have to come up with some very persuasive arguments for me to part with them," as he says. One such specimen is a 1-kilogram green beryl from the Ukraine. In another corner the world's largest cut fire-opal is seen – 'the Chalchiuhtlicue' of 132 carats.

Should a guest feel in need of a drink at the end of the Wild villa tour, the built-in bar at the back of the Gem Room provides a taste of the in-house production of German Riesling from Constantin Wild's vineyards, located on the Rhine. Who knows? Perhaps a celebration of the acquisition of a fabulous gemstone is in order.

Finally, security measures inspired to a great extent by James Bond movies and Blake Edwards' Pink Panther have been installed. Should a visitor accidentally trigger one of them, he or she will never forget the experience. Doors slam closed! Window shutters go down! Lights flash! And the noise level is beyond description. Be sure that no gem leaves the room without the express permission of Constantin Wild.

'CHALCHIUHTLICUE' FIRE OPAL

132 carat

right:
Impressions from the Gem Room.

Edelsteinzimmer
Gem Room

FR 43-4720

DREAMY BACHELORS

Over the past decades, countless articles in magazines and newspapers have featured Constantin Wild, his father, Fritz Wild, and the gemstones of the company. Fritz Wild was quite the playboy, a reputation that made for good coverage in *Ihre Freundin* – a German magazine, the title of which means 'Your Girlfriend'. In an issue from 1955, Fritz Wild and his friend Max Hahn were featured on a two-page spread as two of the most eligible bachelors in Germany at the time. When Fritz Wild's acquisition and restoration of a thousand-year-old castle called Nollig on the Rhine was publicised, huge baskets of love letters arrived from young ladies. Some of them proposed marriage to 'the king of bling', as he was called in the article. One particular lady even took up residence at a local hotel – for as long as she could afford it – in order to attract the attention and conquer the heart of Fritz Wild. Every morning she would throw a bouquet of flowers over the iron gate to the castle. But her efforts went unrequited.

left, clockwise:
Max Hahn and Fritz Wild in Rio de Janeiro.
Fritz Wild at Nollig Castle.
View of Nollig Castle on the Rhine.
Fritz Wild in Egypt.
Fritz Wild's Borgward Hansa Cabriolet in front of Nollig Castle, early 1950s.

GEM ARCHIVES

Since taking over the helm of the company, Constantin Wild has collected numerous images of the most important gemstones he has sold. "Many important gemstones have passed through our hands, but unfortunately, we do not have images of the ones from past generations. So I am documenting some of the most spectacular stones, and I have a vast collection of images of the ones I have sold. I would like to have kept some of the gems, but unfortunately, to make a living, I have had to part with them. When I think about some of the gems I have sold, I become sentimental about the value they would have today, as compared to what I actually got for them," he concludes.

MANDARIN GARNET
40 carat
RED SPINEL
5 carat
BLUE SPINEL
13 carat
more
TOURMALINE
16 carat
CANARY TOURMALIINE
16 carat

DE ▼

Constantin Wild

DIGITAL MEDIA

As the first gemstone merchant worldwide to have a professional website about coloured gemstones, Constantin Wild established his in 1998 as a community and a virtual meeting point for gemstone lovers. For eight years, he was also responsible for building the website of the International Colored Gemstone Association (ICA). For the ICA, he produced approximately forty background fact sheets on various gemstones. "My father and grandfather continually took part in the making of books and sample pages about gemstones for the general knowledge of the public. When authors today write books about gemstones, they usually come to me for specimen samples, photographs and knowledge, as my company contains generations of accumulated information," Constantin Wild explains.

ERNESTO ARTILLO

"THE PUREST AND MOST THOUGHTFUL MINDS ARE THOSE WHICH LOVE COLOUR THE MOST."

John Ruskin,
'The Stones of Venice',
1851-1853

thank you for sharing my love

PHOTO CREDITS

SLIPCASE COVER: red topaz by Jürgen & Hiltrud Cullmann, Lichtblick Fotodesign

A LIFE WITH GEMSTONES: page 6 illustration by Ernesto Artillo; page 10 portrait and pages 20, 22, 23 Thomas Mumbächer; pages 12, 17 Heike Rost; pages 4/5 red topaz and pages 13, 15, 16, 18, 19, 21 all gem photos Archive Constantin Wild

WILD ADVENTURES: pages 24, 60 Gem Room, 71 world map, 55 portrait Heike Rost; all repros pages 27, 28, 29, 32, 33, 34, 35, 36, 37, 41, 43, 44, 45, 46, 47, 48, 49, 52, 53, 56 Thomas Mumbächer; page 28 cutter Archive Hosser; illustration pages 30/31 The Wild Family Tree Thorsten Lönnecker; pages 25, 26, 39, 42, 46, 50, 51, 54, 60, 61, 62, 63, 67, 69, 71, 75 and pages 59, 65, 68, 70, 74 all gem photos Archive Constantin Wild

GLOBAL ESTIMATES: pages 78, 80 peridots and cabochons, 84/85 Heike Rost; pages 81, 83 Thomas Mumbächer; page 80 portrait with crystal and pages 86, 88, 90, 92, 94, 96 all gem photos Archive Constantin Wild

THE MAKING OF COLOUR: pages 98, 100, 104, 106/107 metamorphosis, 108, 109, 110, 112, 113, 114, 116, 117, 120 (background) Thomas Mumbächer; page 118 design drawing and 119 green beryl earrings Stanislav Drokin; background pages 102, 103, 121 Heike Rost; 122 Archive Constantin Wild; all gem photos pages 102, 103, 120, 121, 122, 123 Archive Constantin Wild

THE JOY OF COLOURS: pages 124, 126, 166 Heike Rost; page 126 jewellery drawings Thierry Martin; bottom left Myerson; pages 128, 129, 131, 132, 133, 134, 135, 136, 137, 138, 139, 140, 144, 147, 150, 152, 155, 157, 159, 164, 165, 169 all gem photos Archive Constantin Wild; page 143 brooch from Tiffany & Co. Spring Collection; page 156 rings Paloma Picasso for Tiffany & Co. Holiday Edition; page 148 Cornelia Heinz

OPINION ON: page 170 portrait Katerina Perez by RéanArt Creative Studio; pages 172, 173, 174, 176, 177, 178, 179, 180, 181, 182, 183, 184, 185, 186, 187, 188, 189 all gem photos Archive Constantin Wild

SUPERLATIVE JEWELLERY: page 190 courtesy of Fred Leighton; pages 192, 194, 195, 196 all gem photos Archive Constantin Wild; page 193 Monica Rich Kosann; pages 198, 200, 201 Tiffany & Co.; page 200 portrait Paloma Picasso courtesy of photographer Nigel Parry; page 203 David Webb; page 204 Mikimoto; page 205 Marina B; pages 206, 208, 209 Chopard; page 211 Anna Hu; page 212 Ginza Tanaka; page 213 Suwa; pages 214/215 Pomellato; pages 216, 217, 218, 219 Bulgari; page 221 Temple St. Clair; page 222 Theo Fennell; page 223 Lester Lampert; page 225 Stanislav Drokin; pages 226/227 Oscar Heyman; pages 228/229 Chow Tai Fook; page 230 photographer Tom Wolf for Juwelier Friedrich; page 231 photographer Benjamin Krohn for Kinzel & Rall; pages 232/233 Cindy Chao; page 234 Fred Leighton; page 237 photographer Leo Bieber for Doris Hangartner; page 238 Christie's Images Ltd. 2021

MEN VERSUS WOMEN: page 240 ASjack, stock.adobe.com; pages 241, 242, 243 all gem photos Archive Constantin Wild; page 243 moonstone hippo and tanzanite cufflinks Longmire London

SHARING THE LOVE: pages 244, 247 Heike Rost, 249, 256, 259, 260 all gem photos Archive Constantin Wild; page 250 Bumann, stock.adobe.com; page 253 backdrop wowinside, stock.adobe.com; gem layout Archive Constantin Wild background by Stefan Lohmeyer; page 256 background Archive Constantin Wild; page 257 paraiba ring by Paula Bailey Brent, imperial topaz ring by Cyril Despointes both photos Bob Krasner; page 259 background Rymden, stock.adobe.com; page 260 background Archive Constantin Wild; page 263 Steve Cukrov, stock.adobe.com

GEMSTONES AS LUXURY: page 264 courtesy of Fred Leighton; page 266 background Anton Belovodchenko, stock.adobe.com; page 269 background GraphicCompressor, stock.adobe.com; page 270 background Prostock-Studio, stock.adobe.com; page 273 background Rymden, stock.adobe.com; page 274 background konradbak, stock.adobe.com; pages 266, 269, 270, 273, 274 all gem photos Archive Constantin Wild

THE GEM HOUSE: pages 276, 277, 280, 283, 286 Strong Room, 289 Gem Room and gem bar, 292 background Heike Rost; pages 278, 279, 281, 287, 288 Archive Constantin Wild; page 279 Fotograf Reusse, Idar; pages 282, 283 monitor, 285 repro and mineral, 286 Strong Room, 289 green rough Thomas Mumbächer; page 283 dops, sign, basement Heike Rost, hammer Martin Glauner; page 290, all gem photos 288, 293, 298, 299 Archive Constantin Wild, page 294 kiss konradbak, stock.adobe.com; page 296 illustration by Ernesto Artillo

Special Thanks

In closing this book, I would like to express my very special thanks to my mother, Christel Wild, and my father, Fritz Wild, and to the following companions who have played their part in my work and in getting this book together:

Dirk Allgaier, Ernesto Artillo, Francisco H. Bank, Gareth Bartley, Vivienne Becker, Guy Bedaria, Danielle De Bie, Georg Brandt, Hans Peter Brandt, Wendy Brouwer, Manuel Carrera, Vincenzo Castaldo, Enrica Castelli, Katinka Champion, Cindy Chao, Temple St. Clair, Francis Cui, Hiltrud Cullmann, Jürgen Cullmann, Maria Doulton, Stanislav Drokin, Rudolf Dröschel, Richard R. Drucker, Eli Eliezri, Mark Emanuel, Theo Fennell, Jannis Fostieris, Armin Franzmann, Stephan Friedrich, Werner Fritsch, Annette Fuhr, Sandalio Garcia, Verena Glaese, Cassandra Goad, Thomas Gübelin, Ursel Haggeney, Tobias Hahn, Nina Hald, Mehboob Hamza, Doris Hangartner, Nobuyuki Harada, Kerstin Hendess, Jürgen Henn, Stefan Hey, Lewis Heyman, Tina Hsu, Anna Hu, Willi Hubor, Alfred Huth, Otto Jakob, Ibrahim Jilani, Philipp Kaucher, Marilu Klar, Sonja Klettke, Monica Rich Kosann, Michael Köhler, Jenny Kovach, Oleg Kudelkin, Greg Kwiat, David Lampert, Dr. Thomas Lind, Jörg Lindemann, Stefan Lohmeyer, Thorsten Lönnecker, Rupert Longmire, Edgar Luft, Dr. Çiğdem Lüle, Thierry Martin, Uli Mayer, Bärbel Meyer, Thomas Mumbächer, Bernd Munsteiner, Maximilian Nagy, Heidi Nilges, Katerina Perez, Paloma Picasso, Marina de Ponti, Joachim M. Rall, Werner Reitenbach, William Rosenberg, Gary Roskin, Heike Rost, Ina Roth, Amanda Joy Rubin, Alan Roup, Angela Schell, Shelly Sergent, Yoshiaki Shinohara, Vitória Melo, Andrea Sohne, Pavel Sokolov, Sonja Steffen, Peter Steiniger, Yasukazu Suwa, Jürgen Thom, Cynthia Unninayar, Lothar Wannemacher, Gernot Weber, Klaus Eberhard Wild, Tommy Wu

... as well as a big thank you to all the family members, faithful employees, gem cutters, suppliers, loyal customers, friends and business associates who have worked alongside us and given us their support during the past 175 years.

Idar-Oberstein, May 2021

Constantin Wild

GEMS, COLOURS & WILD STORIES
175 YEARS OF CONSTANTIN WILD

Constantin Wild / Nina Hald (eds.)

Authors
Vivienne Becker, Katinka Champion, Maria Doulton, Richard R. Drucker, Nina Hald, Çiğdem Lüle, Katerina Perez, Gary Roskin

Translations and editing
Gareth Bartley

Copy editing
Wendy Brouwer

Graphic design
Lifestudio, Thorsten Lönnecker

Offset reproductions
Schwabenrepro, Fellbach, Germany

Printed and bound by
Nino Druck, Neustadt/Weinstrasse, Germany

Paper
150 g/m2 Luxoart Samt

Bibliographic information published by the Deutsche Nationalbibliothek
The Deutsche Nationalbibliothek lists this publication in the Deutsche Nationalbibliografie; detailed bibliographic data are available on the Internet at www.dnb.de

ISBN 978-3-89790-626-6

Made in Germany, 2021

www.constantinwild.com

Constantin Wild GmbH & Co. KG
Hauptstrasse 103
55743 Idar-Oberstein
Germany

www.arnoldsche.com

arnoldsche Art Publishers
Olgastrasse 137
70180 Stuttgart
Germany